Introduction

It is the intention of the authors that this dictionary should assist students of general engineering design and manufacture in their studies. It is suitable for students at various levels of study up to and including higher technicians, both BTEC and graduate.

The length of any dictionary is determined by commercial considerations and this one is no exception. Therefore, the authors have endeavoured to give as broad a coverage as possible within reasonable limits of size and cost. To this end, four main topic areas have been considered and the number of definitions has been shared almost equally between these topic areas.

- mathematics and statistics,
- engineering science,
- electrical engineering, electronic engineering and information technology,
- mechanical engineering, manufacture, computer-aided engineering and materials.

In addition, there are a number of appendices covering such topic areas as Laplace transforms and standard forms for derivatives and integrals, together with tables of SI units and conversion factors.

The definition of many engineering terms depends on the context in which the term is used. For example, a *filter* is a device for removing foreign matter from an oil when considering lubrication, but a *filter* can also be an electrical circuit for accepting or rejecting a signal of a particular frequency or band of frequencies. Therefore, to avoid confusion, each headword (in bold type) is followed by a 'key' in parentheses to indicate the context in which that word is being defined. The keys are as follows:

- Mathematics — (*maths*)
- Statistics — (*stats*)
- Engineering science — (*sci*)
- Electrical and electronic engineering — (*elec*)
- Information technology — (*comp*)
- Mechanical engineering and manufacture — (*mech*)
- Computer-aided engineering — (*CAE*)
- Materials and heat treatment — (*matls*)

For example:
abrasive (*matls*) Hard materials such as aluminium oxide or silicon carbide, usually in powdered form, used for grinding or abrasive machining processes.

Finally, although every effort has been made to try and select topic areas that you will find useful, inevitably different users will have differing requirements. Therefore the authors and the publishers would appreciate constructive suggestions for additions, deletions and corrections.

R.L. Timings
P. Twigg

A

abradant (*matls*) Any abrasive material used for grinding, honing, lapping or polishing processes.
abrasive (*matls*) Hard materials such as aluminium oxide or silicon carbide, usually in powdered form, used for grinding or abrasive machining processes.
abrasive wheel (*mech*) A grinding wheel used for cutting and finishing materials; comprising an *abrasive* material and a bonding agent formed in the shape of a wheel that can rotate about an axis for material removal; generally for use with grinding machines and power tools.
absolute address (*comp*) Address of memory location of a computer specified as a fixed number rather than relative to the program counter content. See: *relative address*.
absolute pressure (*sci*) Pressure measured with respect to zero pressure. Most pressure gauges, such as Bourdon tube types or manometers, measure gauge pressure, p_g, i.e. pressure above the atmospheric pressure. To obtain the absolute pressure, p_a, atmospheric pressure, p_{atm} must be added to the gauge pressure; thus

$$p_a = p_{atm} + p_g$$

absolute temperature (*sci*) Also known as the *thermodynamic* temperature scale. Temperature measured with respect to *absolute zero*.
absolute zero (*sci*) The temperature at which the kinetic energy of all molecules and atoms ceases to exist, absolute zero measured on the Kelvin scale (0 K) = –273.15°C. For most engineering

purposes, to convert degrees Celsius to Kelvin add 273°C; for example,

$$T(\mathrm{K}) = T(°\mathrm{C}) + 273$$

a.c. (*elec*) See: *alternating current.*
acceleration (*sci*) The rate of change of velocity, quantity symbol a, measured in m/s^2, a vector quantity like velocity. The change in velocity can be a change in magnitude, such as a body increasing speed in a straight line, or a change in direction, such as vehicle cornering at a constant speed. See also: *retardation*; *centripetal force*.
acceleration due to gravity (*sci*) The acceleration of a body falling freely in the Earth's gravitational field. Air resistance affects the acceleration so it is experimentally measured in a vacuum. The value of g (gravitational acceleration) varies slightly over the earth's surface as it varies with the distance to the earth's centre and the earth is not spherical. For engineering purposes a value of 9.81 m/s^2 is generally accepted. This can be used to calculate the weight of a body from the mass. Applying Newton's second law, since force equals mass multiplied by acceleration ($F = ma$), then weight $= mg$, where weight is a special case of force. The attractive force between two bodies is dependent on the mass of each; for example, the mass of the moon is significantly less than the Earth and consequently has a much lower value of $g(g = 1.624 \text{ m/s}^2)$.
accumulator 1 (*comp*) Computer register where the solution to an arithmetic operation or data transfer is stored. **2** (*elec*) Electrical cell or battery (also referred to as a *secondary* cell or battery) that can be charged and discharged, useful for storing electrical energy in a portable form, e.g. a vehicle battery.
accuracy 1 (*sci*) A measure of how close the desired output of a process is to the actual mean output. For example, in instrumentation accuracy describes how close the measured value of a quantity is to the actual value. This is usually expressed as

a maximum error, e.g. ± 1 mV, or as a percentage of the full range of the instrument, e.g. 0.5%. Not to be confused with *precision.* **2** (*mech*) A measure of how close the actual size of a component dimension approaches the size specified by the designer. The dimensional tolerance specified by the designer.
acme thread (*mech*) Thread having a profile with an included flank angle of 29° and a flat crest and a root; it is generally used for transmitting motion in conjunction with a disengaging nut such as for the lead screw on a lathe. The tapered sides of the thread profile eases engagement of the screw and nut, and provides automatic adjustment as wear takes place, owing to deeper engagement of the screw and nut (see: *screw thread*).
a.c. motor (*elec*) Electrical motor that operates from an alternating current supply; see also: *synchronous motor*; *induction motor*.
actuator (*mech*) Any device used in power transmission systems to convert the energy of the transmitting medium into mechanical energy and movement by providing the required controlled action; for example, a hydraulic cylinder converts the pressure energy of the fluid into linear thrust. Common types are pneumatic, hydraulic and electric. Control of the actuator may be manual or automatic.
acute-angled triangle (*maths*) See: *triangle.*
addendum (*mech*) Distance between the crest of a *gear* tooth and its *pitch circle* measured radially.
adder (*comp*) Electronic device that adds *binary* numbers.
addition law of probability (*stats*) This law is recognized by the use of the word '*or*' connecting the probabilities. Thus if p_X is the probability of event X happening and p_Y is the probability of event Y happening, the probability of events X *or* Y happening is $p_X + p_Y$. This law can be used for any number of events connected by the word '*or*'.
address (*comp*) Code for data or instructions to be located in a computer memory. See: *absolute address*; *relative address*.

address bus (*comp*) A set of parallel wires used to specify the address for data transfers to or from the microprocessor of a computer.
adherend (*matls*) Any materials joined by an adhesive.
adhesion (*matls*) **1** The bonding together of matter by intermolecular forces. **2** The joining together of two surfaces by use of an adhesive, in which the intermolecular forces between the adhesive and the adherend provide the bond. **3** The intimate bonding together of metallic surfaces resulting from compressive stresses, and as a function of time and temperature (see: *cold welding*). **4** (*elec*) The mutual forces holding together two magnetized bodies linked by magnetic flux fields, or the mutual forces holding together two electrostatically charged particles.
adhesive (*matls*) A substance for joining two materials together by adhesion. In engineering applications the adhesive is usually a synthetic polymer. The adhesion depends on the attractive forces between the molecules in the surface and simple mechanical keying between the adhesive and surface irregularities of the adherend. The adhesive must be able to mould perfectly to the contours of the surfaces to be joined to ensure close molecular contact. Thermosetting resins such as epoxides have a high viscosity and are useful as high-strength, gap-filling adhesives. Thermosetting resins such as cyanoacrylates have a low viscosity and are particularly useful where surface penetration is required. Thermoplastic adhesives such as polystyrene cements are useful for joining non-metals. They are gap-filling and flexible.
adhesiveness (*matls*) A foundry term describing the ability of particles of sand to cling to another material.
adiabatic process (*sci*) Any thermodynamic process occurring without any heat transfer across the system boundary.
admiralty brass (*matls*) A brass alloy consisting of 70% copper, 29% zinc, 1% tin and 0.01 to 0.05% arsenic. It is a ductile alloy,

similar in properties to cartridge brass but having improved corrosion resistance particularly in the presence of sea water as its name suggests.

admiralty gun metal (*matls*) See: *bronze*.

admittance (*elec*) The property of an electrical circuit which allows a flow of current in the circuit when a potential is applied across the circuit. Admittance, the reciprocal of *impedance*, has the symbol *Y*. The units are Siemens (S).

aerodynamics (*sci*) A branch of fluid mechanics that concerns the dynamics of gases. Usually the term is used to refer to the study of forces acting on a body in motion in air.

aerofoil (*sci*) A device designed to produce a normal reaction to the direction of motion, e.g. an aeroplane wing providing lift. The shape must also produce minimum drag in the direction of motion.

age-hardening (*matls*) A structural change that takes place spontaneously in some metallic alloys resulting in a change in the mechanical properties of the alloy, usually causing an increase in strength and hardness with a corresponding loss of ductility. See: *precipitation hardening*.

air bearing (*mech*) A frictionless bearing in which the shaft and bearing shell are separated by high-pressure air. Figure A.1 shows that the bearing shell is perforated with fine holes through which air is forced under high pressure. There is an appreciable gap between the shaft and the bearing shell and the shaft floats on a cushion of air. Normally the air pressure around the shaft is uniform; however, if the load on the shaft causes it to move off-centre the resulting imbalance in pressure moves or tends to move the shaft back to its central position. Therefore, the system is self-centring. For heavier duty applications pressurized oil can be used instead of air and the bearing is known as a hydrostatic *bearing*.

air capacitor (*elec*) A capacitor that uses air as its dielectric, most commonly used as a variable capacitor for tuning purposes; one set of plates are fixed and the other is rotated to vary the overlap and capacitance.

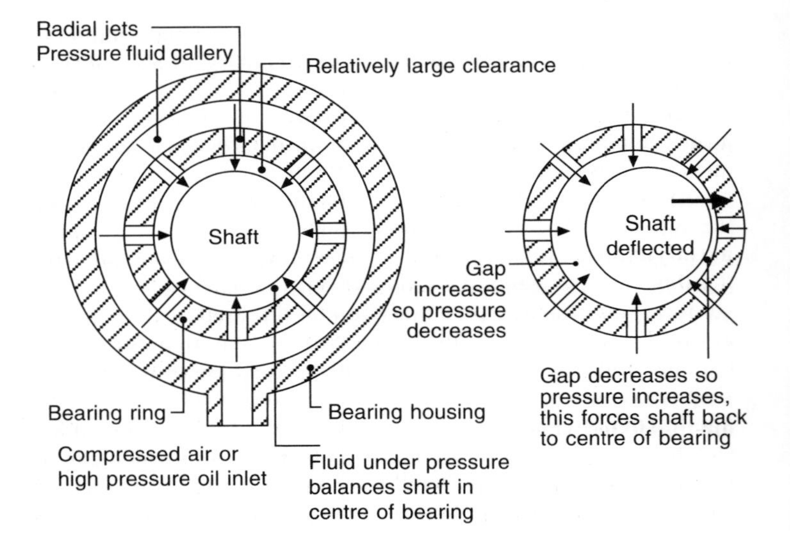

Fig. A.1 Air Bearing

air-chuck (*mech*) A work-holding device for use on automatic lathes and both CNC lathes and milling machines, which produces a clamping action using compressed air. This provides a more constant clamping pressure and more rapid loading of workpieces than a manual chuck. In a diaphragm chuck the air pressure provides the clamping force and, in the event of a pressure failure, the work is released. In a toggle chuck the air pressure only operates the toggles and, in the event of a pressure failure, the work remains securely held, i.e. the chuck is 'fail-safe'.
air compressor (*mech*) A machine that takes air at atmospheric pressure and compresses it to a higher pressure, commonly for use in power transmission or control systems, the mechanism of the compressor being either reciprocating or rotary.

air ejector (*mech*) A device for maintaining a partial vacuum in a vessel such as a condenser. The ejector consists of a jet of fluid that passes through a diverging nozzle to reduce the pressure, thus drawing air and vapour from the vessel through a pipe before passing the working fluid together with the air and vapour to a diffuser and the outlet.
air filter (*mech*) Cleaning device fitted to the intake of internal combustion engines or air compressors to remove dust particles. Large quantities of air are ingested and air filters prevent abrasive particles from entering the unit and causing damage.
air-hammer (*mech*) A forging hammer powered by compressed air that may be controlled manually or automatically. The compressed air allows sensitive control and the force of the blows may easily be varied.
air-hardened steel (*matl*) A steel alloy, typically containing 0.34% carbon, 4.25% nickel and 1.25% chromium, which can be hardened by heating it to 830°C and cooling it rapidly (quenching it) in an air blast. The alloy is then tempered. Air-hardened steels are used for small, intricate components requiring a high strength but minimum distortion during quenching.
airlock (*mech*) An air pocket in a fluid system that prevents the flow of liquid.
air receiver (*mech*) A pressure vessel for the storage of compressed air. As well as acting as a reservoir, the receiver not only smoothes out any pressure fluctuation in the system but also allows the removal of moisture which must be drained from the receiver from time to time.
air-standard cycles (*sci*) Equivalent gas cycles using air as the working fluid for assessing the performance of internal combustion engines, the cycle efficiencies being called *air standard efficiencies*. The three air-standard cycles which describe the performance of reciprocating i.c. engines are the *Otto cycle* (basis of the spark ignition engine), the *Diesel cycle* and the *mixed* or *dual cycle*.

algebra (*maths*) A system used in mathematics to represent variable quantities by symbols for investigation. Expressions usually take the form of equations which may be rearranged and solved. Also applied to the study of series, sets and matrices.
algebraic sum (*maths*) The addition of a set of quantities paying attention to the sign, e.g.

$$2x - 4x - 3x = -5x$$

ALGOL Abb. (*comp*) Algorithmic Language; one of the first high-level languages with structured programming, which has now almost completely been replaced by *Pascal.*
algorithm (*comp*) A set of precise and unambiguous rules which specify a sequence of actions. An algorithm is defined in preparation for the programming of a computer.
aliasing (*stats/elec*) Frequency distortion when sampling a data-sampled or discrete system which occurs with all data frequencies above the *Nyquist frequency*, i.e. signal frequencies which are above half the sampling frequency. When sampling a signal, components at frequencies higher than the Nyquist frequency appear identical to components at a lower frequency.
alloy (*matl*) **1** A mixture of miscible metals and/or non-metals added to a base (solvent) metal in the molten state. **2** A mixture of metal and/or non-metal powders added to a base metal powder and sintered. In the solid state the alloy consists of compounds, solid solutions and/or eutectic mixtures. The purpose of forming alloys is to produce new metals with special properties such as high strength, corrosion resistance, heat resistance, improved hardness and improved fatigue resistance.
alloy cast iron (*matl*) These contain appreciable amounts of nickel, chromium, copper, vanadium and molybdenum to enhance the properties of the alloy. A nickel alloy (Ni-cast) is widely used for machine frames as it produces fine-grained castings that are wear resistant and are dimensionally stable. Increasing the nickel

content to between 4% and 6% and adding up to 1% chromium produces hard, wear-resistant alloy (Ni-hard) that is naturally martensitic as cast, without the need for heat treatment. This alloy can only be machined by grinding. Increasing the nickel content still further to between 11% and 20% together with up to 5% chromium produces an alloy that is naturally austenitic at room temperature. Such cast irons are corrosion-resistant, heat-resistant, tough and non-magnetic.

alloy steels (*matl*) These are carbon steels that contain various significant amounts of alloying elements such as nickel, chromium, copper, manganese, molybdenum, vanadium, tungsten and cobalt, depending on the properties required. Manganese steels and nickel–chromium steels are strong and tough. Steels containing tungsten and cobalt retain their hardness at high temperatures and are used for cutting tools. Steels containing high percentages of chromium are corrosion resistant.

alphanumeric (*comp*) Set of characters on a computer keyboard consisting of alphabetic characters (A to Z) and numeric characters (digits 0 to 9) which are represented by binary numbers for processing by a computer, e.g. *ASCII* is a common alphanumeric code.

alternating current (a.c.) (*elec*) A bidirectional electric current that flows alternately in opposite directions around a circuit with a constant frequency, unlike direct current which is unidirectional. This is the type of supply obtained from the mains at a frequency of 50 Hz, i.e. it changes direction 100 times per second. Most alternating currents have a sinusoidal waveform and this allows simple algebraic study. Alternating currents are the product of the *alternator* and *electronic oscillator*.

alternator (*elec*) An electrical generator consisting of either coils that rotate in a magnetic field or electromagnets rotating inside fixed coils. This type of generator has no *commutator* and it produces an alternating current whose frequency depends on the speed of the rotor and the number of magnetic pole pairs.

aluminium (*matl*) silver–white metallic element, density 2710 kg/m^3, melting point 932 K. It is a good electrical conductor and an oxide layer that forms in air provides good corrosion resistance. Pure aluminium is too soft and weak for most engineering purposes and so alloys of aluminium are commonly used for sheet, sections and castings. Because of their lightness and good electrical conductivity, pure aluminium conductors (with a steel core to withstand the mechanical loads) are used for the overhead electrical cables of the national grid system.
ammeter (*elec*) An instrument for measuring the flow of current in an electric circuit. An ammeter must be connected into the circuit in series so that the current flows through the meter. Therefore the ammeter must be capable of carrying the maximum current liable to flow in that circuit.
ampere (amp) (*elec*) The SI unit of electrical *current*. It is defined as the current which, if maintained in two parallel conductors of infinite length, of negligible cross-section, and placed one metre apart in a vacuum, would produce a force between the conductors of 2×10^{-7} N m^{-1}. One ampere is equal to: **1** the flow of one coulomb of charge per second around a closed circuit. **2** the flow of 6.28×10^{18} electrons per second passed any point in a closed circuit. Unit symbol A.
ampere–hour (*elec*) Unit of electrical energy or charge used for measuring the capacity of accumulators. For example, a 40 Ah battery could supply 4 amps for 10 hours before the voltage falls below a given value. It is also the unit for measuring the rate of energy consumption of a domestic supply. It is equivalent to 1 *ampere* of current flowing for 1 *hour* = 1 Ah.
ampere–turn (*elec*) SI unit of *magnetomotive force*, symbol At, derived from the effect of a *current* flowing through a *conductor* forming a coil or *solenoid*. The magnetomotive force, in ampere–turns, equals the current flowing through a coil multiplied by the number of turns of wire making up that coil.
amplifier (*elec*) A device for increasing the strength of a signal

without significantly altering its characteristics, operating by drawing energy from an external source. It may be made from discrete components or a single *integrated circuit* may be used.

amplitude (*elec*) Maximum displacement of a periodic varying quantity from the mean value, e.g. the maximum value of a sine wave during any half-cycle.

analogue filter (*elec*) A filter for analogue signals, such as voltage or current that vary continuously with time. Passive electronic analogue filters usually consist of combinations of resistors, capacitors and inductors designed to accept certain frequencies and reject (filter out) all other frequencies. Active filters also contain an amplifying element such as a transistor or op-amp.

analogue signal (*elec*) Type of continuously variable signal that may take on an infinite number of values as it mimes the characteristics of the information being transmitted, e.g. potential and current.

analogue-to-digital conversion (ADC) (*comp*) Device for converting an analogue signal into a digital one for computer analysis or transmission.

analogy (*elec*) Corresponding relationships between the quantities at the output and input of electrical systems and other types of system; most systems can be represented by an electronic circuit comprising a network of capacitors, resistors and inductors.

AND gate (*elec*) A *logic gate* which produces an output of logic 1 only when all inputs are logic 1. The MIL and BSI symbols for an AND gate is shown in Fig. A.2(a). The *truth table* for a two-input AND gate (Boolean expression (A·B) is shown in Fig. A.2(b).

aneroid barometer (*sci*) Type of barometer consisting of hollow bellows made of thin metal, the pressure inside being a partial vacuum. One diaphragm is fixed and the movement of the other, which reflects changing atmospheric pressure, is amplified via a train of levers to a scale pointer or to a pen chart.

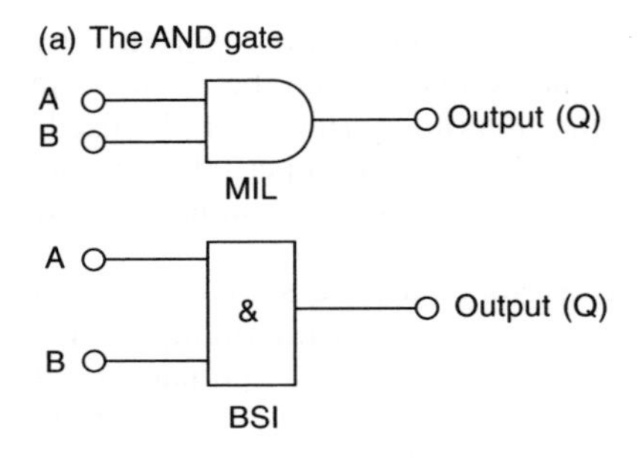

(b) AND gate truth table

A	B	Q
0	0	0
0	1	0
1	0	0
1	1	1

Fig. A.2 AND Gate

angle (*maths*) The inclination of one line to another, measured in degrees of arc or radians. Any angle whose value lies 0° and 90° is called an *acute angle*. An angle of 90° is called a *right angle*. Any angle whose value lies between 90° and 180° is called an *obtuse angle*. Any angle whose value lies between 180° and 360° is called a *reflex angle*. Any two angles whose values add up to 90° are said to be *complementary* and any two angles whose values add up to 180° are said to be *supplementary*.
angle cutter (*mech*) A cutter for use with a horizontal spindle milling machine. It is used to cut a flat surface at an angle to the cutter axis. See: *milling cutters*.
angle Dekkor (*mech*) This is not a direct-reading instrument but is an optical instrument for measuring angles by comparison. An initial reading is taken over combination angle gauges selected to give the required angle. The combination angle gauges are replaced by the component to be checked and a second reading is taken. Any deviation from the correct angle can be determined by calculating the difference between the two readings. See: *autocollimator*; *collimation*.
angle iron (*matl*) A rolled steel L-section bar used for the

fabrication of strong angle joints. Each face of the angle iron is riveted or welded to each side of the plates to be joined. It is also used in the construction industry for connecting girders at a right angle to each other. It is left in a hot-rolled finish in the larger sizes but is available in a cold-drawn finish in the smaller sizes.

angle of friction (*sci*) The angle between the normal reaction of a surface acting on a body, and the resultant of the frictional force and the normal reaction. This angle, ϕ, may be calculated from the coefficient of friction, μ, as follows: $\mu = \tan \phi$

angle of twist (*sci*) Angle through which one cross-section of a shaft subject to a torque turns (twists) relative to another section; a significant quantity for torsion equations measured in radians.

angle plate (*mech*) An L-section casting having slots for the attachment of clamps and accurately machined (and sometimes ground) to provide surfaces that are mutually and accurately perpendicular to each other. **1** Angle plates are used in conjunction with a surface plate or a marking out table for marking out work prior to machining. **2** Angle plates are used to support and locate workpieces perpendicular to the work table surfaces of machine tools such as milling machines.

angular acceleration (*sci*) Rate of change of angular velocity, commonly measured in rad/s^2 for engineering purposes.

angular displacement (*maths*) The angle turned through by a body about an axis, measured in radians or degrees of arc.

annealing (*matl*) A heat-treatment process for making a metal or alloy as soft as possible prior to a flow-forming operation and to relieve internal stresses. (See: *normalizing.*) In the case of ferrous metal and alloys, the metal is heated to above the upper critical temperature for hypo-eutectoid steels and to above the lower critical temperature for hyper-eutectoid steels dependent upon the carbon content of the steel as shown in Fig. C.12 and then allowed to cool as slowly as possible; this is best done by heating in a furnace and then turning the furnace off to cool with the metal still inside. In the case of non-ferrous metals, the metal is heated to just above its temperature of *recrystallization*. See: *recrystallization*.

annular gear (*mech*) A gear in the form of an annulus with the teeth cut into the internal periphery to engage with a pinion.

anode (*elec*) A positively charged electrode in any electrical or electronic device to which negatively charged particles (*electrons*) or negatively charged ions (*anions*) are attracted.

anodic protection (*matl*) Corrosion protection system using replaceable, sacrificial anodes to protect a metal structure from corrosive attack. For example, replaceable blocks of zinc are bolted to the hull of a ship in the vicinity of the manganese bronze propellers. The manganese bronze and the low-carbon steel hull plating form an electrolytic cell with the highly agitated sea water as the electrolyte. Without anodic protection the hull would be quickly eaten away. However, the zinc blocks are anodic relative to the steel and the manganese bronze and are gradually eaten away whilst protecting the propellers and the steel hull.

anodizing (*matl*) The surfaces of aluminium and its alloys develop a protective coating of aluminium oxide in the presence of atmospheric oxygen. This oxide coating can be made thicker and built up more quickly by an electrolytic process called *anodizing*. The anodized oxide layer is produced by electrolytic action where the aluminium is made the *anode* in an oxidizing *electrolyte*. The naturally occurring oxide is grey in colour but the anodized oxide film can be coloured more attractively by the use of selected acids for the electrolyte or by the use of dyes after anodizing.

anomalous viscosity (*sci*) The description of liquids whose *viscosity* reduces as the rate of flow or velocity gradient increases.

anti-friction bearings (*mech*) See: *ball-bearing*; *roller-bearing*; *air-bearing*.

anvil (*mech*) A heavy iron support on which work is formed during forging, traditionally used by a blacksmith. The bed of a power-hammer that supports the lower die or tool.

aperiodic (*elec*) Classification of deterministic signals which are not repetitive, sometimes called transient signals.

apparent power (*elec*) The product of the voltage and the current in an a.c. circuit as measured by a voltmeter and an ammeter, as opposed to the *true power* measured by a wattmeter which allows for phase shift in reactive circuits.

apron (*mech*) A component of a lathe attached to the saddle (combined assembly of carriage and apron) and which carries the mechanism, together with the appropriate controls, that provides the sliding, surfacing and screw-cutting motion of the saddle as it moves along the *bed*. See: *centre lathe*.

arbor (*mech*) A shaft upon which cutters may be mounted in a horizontal milling machine. One end of the arbor is located in, and driven by, the machine spindle, the outer end being supported in a steady bearing. See: *stub arbor*.

Archimedes' Principle (*sci*) A body immersed partly or wholly in a fluid experiences an *upthrust* equal to the weight of the displaced fluid. The weight of the fluid displaced is equal to the weight of the body. A body will float in a fluid only if it is able to disperse sufficient weight of the fluid equal to its own weight.

arcing contact (*elec*) Component of circuit breaker which receives most of the damage from arcing when opening and closing the circuit, and which is designed for ease of replacement.

arcing voltage (*elec*) **1** The voltage between two electrodes that is necessary to cause the dielectric separating the electrodes to break down so that a current is able to flow. **2** The total voltage across an electric arc, usually referred to when *electric arc welding*.

arc voltage (*elec/mech*) The potential difference (voltage) across an electric arc, usually referred to when electric arc welding. See: *electric arc welding*.

arithmetic (*maths*) The science of numbers including such processes as addition, subtraction, multiplication, division, and the determination of powers and roots.

arithmetic logic unit (*comp*) Programmable computer device which can perform different mathematical functions on binary words applied to its inputs.

arithmetic mean (*stats*) See: *mean*.

arithmetic register (*comp*) Temporary memory location of computer used to hold results during mathematical processing.

arithmetic series (*maths*) A series of numbers with a fixed difference between terms, e.g. 1, 4, 7, 10, 13, ... is an arithmetic series with a progressive difference of +3. The formula for an arithmetic series is usually expressed for the *n*th term as $a + (n - 1)b$ where a is the first term in the series and b is the difference between terms. See also: *geometric series*.

armature coil (*elec*) A conducting coil of the *armature* of electric machines that produces a magnetic flux that interacts with the flux of the field magnet so as to produce rotary motion in an electric motor, or so as to produce an induced electric current in a generator.

array 1 (*maths*) The numbers within a matrix. See: *determinants*, *matrix*. **2** (*comp*) A series of data of the same type stored in successive storage locations, where each element is referenced by a numerical *address*. In data analysis, a set of sampled data of a variable is formed in an array prior to processing. **3** (*elec*) In telecommunications it can refer to an antenna system.

artificial intelligence (*comp*) Computer science opines that computers can behave in ways considered to be similar to human intelligence and demonstrate such characteristics as reasoning, learning and adaptive behaviour. Certain *fuzzy logic* schools of thought consider that artificial intelligence only refers to bivalent techniques such as expert systems and not to neural network and neuro-fuzzy reasoning techniques.

ASCII Abb. (*comp*) Abb. of American Standard Code for Information Interchange. A numerical code that allows textual information to be processed and stored by computer systems, each individual key of a keyboard being represented by a unique 7-bit binary code. The ASCII code is now used by virtually all computer manufacturers and allows textual information to be understood by different computer systems.

aspiration (*sci*) The induction of air into an internal combustion engine. A naturally aspirated engine induces a fresh charge by the pressure reduction in the cylinder as the piston descends. Compare: *supercharging*.

assembly language (*comp*) Low-level programming language using meaningful mnemonics to represent binary instructions, making programming easier. Conversion of the assembly language into *machine code* is carried out by an assembler program.

associative memory (*comp*) Type of computer memory located by the content rather than an *address*.

asynchronous (*comp*) Process not controlled by an external clock.

asynchronous computer (*comp*) Computer that operates without the control of a master clock; instead the start of one operation is triggered by the completion of the previous one.

atmospheric pressure (*sci*) The pressure exerted by the atmosphere at the surface of the Earth. The standard value is 1.01325×10^5 N/m^2. The pressure decreases with increasing altitude. See: *standard atmosphere*; *gauge pressure; absolute pressure*.

atomic clock (*sci*) Highly accurate instrument, controlled by the frequency of atoms of molecules, used for the accurate time measurement.

attenuation (*elec*) The reduction in the amplitude of a signal, i.e. the opposite of *amplification*.

audio frequency (*elec*) Any frequency that is audible to a human ear, generally in the range of 20 Hz to 20 kHz.

austenite (*matl*) A solid solution of carbon in iron that is present in plain carbon steels above certain critical temperatures (see: *iron–carbon equilibrium diagram*, Fig. C.12). As for all solid solutions of metals, steel in the austenitic condition is soft and malleable and, consequently, it is more easily shaped by hot-working (e.g. hot-rolling and hot-forging). Austenite is also non-magnetic and magnetic tests on the steel are a useful method for determining the upper temperature of the *critical range*. Although austenite cannot exist at room temperature in plain carbon steels, it can exist at room temperature in some alloy steels, notably 18/8 stainless steel.

autocollimator (*mech*) This is a direct reading instrument for angular measurement. It is used for checking machine tool alignments and similar applications. It projects an image of its target wires onto a reflector using collimated light. The reflected image is returned through the collimating lens system so that both the target wires and their image are visible when viewed simultaneously through a magnifying eyepiece. The target wires (setting lines) are adjusted by a micrometer control until they straddle the reflected image. The reading is then taken as the nearest minute of arc of the graticule scale in the eyepiece plus the micrometer reading. The divisions of the micrometer scale each represent 0.5 of a second of arc. If there is no error, the reflected image would be superimposed on the target wires without any adjustment having to be made. See: *angle Dekkor*; *collimation*.

auto-correlation (*elec*) Measure of the dependence of a signal at one instance in time on itself at another instance in time. See also: *cross-correlation*.

automatic control systems (*elec*) **1** System for controlling the value of a plant output quantity at a desired value. Feedback provides information to the control system about the actual quantity value, and as external disturbances cause deviations in the quantity, the control effort to the plant input is adjusted automatically to attempt to maintain the correct output value. **2** A sequential system

involving control switches operated at the correct time and in the correct sequence. Sensors provide on/off signals about the state of the plant, and logic control functions are used to operate the plant actuators.
automatic lathe (*mech*) A lathe used for the mass production of turned parts. The functions of starting, stopping, bar feed and tool movements are controlled by cams which are changed for each job by the machine setter. Small lathes have a single spindle but larger lathes have several spindles (multi-spindle) all but one operating at the same time whilst the stationary spindle is being loaded and unloaded.
avalanche current (*elec*) Under normal operating conditions the depletion layer at the pn junction of a diode is devoid of charge carriers and forms a region of high resistivity. If the reverse voltage is increased to such an extent that it causes the depletion layer of diode to break down, the minority charge carriers travelling through the depletion layer dislodge valency electrons. This effect is cumulative causing a build up of reverse current flow (*avalanche current*) that eventually destroys the diode. See: *zener diode.*
average (*maths*) The arithmetic average, or mean, of a set of *n* numbers is the sum of the numbers divided by *n*. See: *modal value*; *mean*, *median.*
axial flow compressor (*mech*) A multi-stage compressor comprised of alternate stages of rotating blades driven by a rotor and stationary blades attached to the compressor casing, causing the air to be forced axially through a decreasing area of flow as the pressure increases.
axial pitch (*mech*) Distance from one point on a screw thread to the corresponding point on the next thread measured parallel to the axis of the thread. Not to be confused with lead. See: *lead.*
axis 1 (*maths*) Mutually perpendicular reference lines on a graph accommodating coordinates values. The *origin* of the graph being the point where the axes cross. **2** (*mech*) A line having a particular

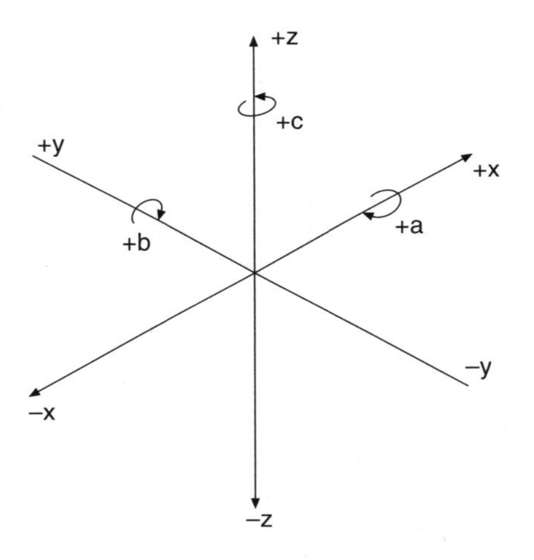

Fig. A.3 Axis Nomenclature

importance and relationship to a problem or system under investigation, e.g. axis about which a body rotates, axis of symmetry.

axis nomenclature (*CAE*) BS3635 provides axis and motion nomenclature for computer-controlled machine tools to simplify programming. The *z*-axis is always the main spindle axis and is always *negative* towards the work for safety—omission of the sign will retract the tool rather than running it into the work and causing an accident. The *x*-axis is always horizontal and parallel to the working surface. The *y*-axis is perpendicular to both the *x*- and *z*-axes. These axes and rotation about these axes are shown in Fig. A.3. SAFETY: Before programming, check the machine manual in case of deviation from this standard.

axle (*mech*) Any shaft which transmits power, supports, or supports and drives, rotating bodies such as wheels.

B

Babbitt metals (*matl*) A 'white metal' plain bearing alloy named after Isaac Babbitt (nineteenth century British engineer), and consists mainly of tin together with antimony, copper and lead in different proportions depending on the service requirements. A relatively weak material, it is used to provide a surface coating on copper bronze bearing shells. Tin and antimony form hard, wear-resistant, low-friction cuboid crystals of a tin–antimony intermetallic compound dispersed throughout a soft matrix of a tin–antimony solid solution. To prevent segregation of the tin–antimony cuboids, which tend to float to the surface because of their lower relative density than the matrix, copper is added to form a copper–tin, intermetallic compound. This copper–tin compound forms before the tin–antimony compound and enmeshes the tin–antimony cuboids as they form and thus prevents any tendency to segregation.
back e.m.f. (*elec*) An electromotive force that opposes the main (forward) current flow in an electrical circuit. For example, when the armature of an electric motor rotates, an opposing e.m.f. is induced in the *armature coils* due to these coils cutting the magnetic flux of the field magnets and this *back e.m.f.* reduces the forward current flow.
backing sand (*mech*) Sand used to fill the moulding box for casting after the facing sand has been rammed around the pattern. Backing sand is sand that has been used previously and, by recycling it, there is a considerable saving in cost.
baffle (*mech*) Device used to redirect or impede the flow of a fluid.
balanced draught (*mech*) System of air supply to a furnace that ensures that the pressure inside the furnace is maintained at

atmospheric pressure; this is achieved by the use of two fans, one to supply air to the furnace grate and the other to exhaust the combustion gases.

ball-bearing (*mech*) Bearing comprised of a series of hard steel balls held between an inner ball-race mounted on the shaft and an outer ball-race held by the casing. The balls are maintained at a constant spacing by means of a *cage*. Such bearings may be journal bearings resisting radial loads or they may be thrust bearings resisting axial loads.

ball-pein hammer (*mech*) Engineer's hammer, the head of which has a flat surface at one end and a hemispherical surface (ball-pein) at the other, used for hand riveting.

ball-valve (*mech*) Valve consisting of a ball resting on a seat allowing fluid to pass in one direction only when the direction of the fluid flow lifts the ball off its seating.

bandwidth (*elec*) The range of signal frequencies that a system will accept for a given purpose without alteration of amplitude or phase; often this is defined as the range that a device can operate in limits specified by a 3dB change from the mean level; this is known as the cut-off frequency.

bar charts (*maths*) The representation of statistical data by equally spaced parallel rectangles. The bars may be horizontal or vertical and the quantity value is represented by the length or height of the bars. Vertical bar charts are frequently called *histograms* which may not be strictly correct. See: *histograms*.

barometer (*sci*) An instrument for measuring *atmospheric pressure*. The simplest type is the mercury barometer consisting of a reservoir of mercury and a glass tube sealed at one end. The tube is filled with mercury and the open end inverted in the reservoir of mercury. The pressure of the atmosphere acts on the surface of the mercury in the reservoir and pushes the mercury up the tube, the force exerted by the atmosphere being balanced by the weight of the mercury column. Atmospheric pressure is then directly proportional to the height of the column and independent of the column cross-sectional area. The relationship is $P = \rho gh$ where P is the atmospheric pressure in N/m^2, ρ is the density of the mercury (or any fluid) in kg/m^3 and g is the acceleration due to gravity in m/s^2. Standard atmospheric pressure is 1.01325×10^5 N/m^2. The *aneroid barometer* is less accurate than the mercury barometer but more practical and convenient for many purposes.

barometric corrections (*sci*) Adjustments that must be made to a barometer reading, e.g. to correct for altitude.

base (*maths*) The basis of a denominational number system. Normally this is 10. However, computers work on a binary system that has base 2. See: *hexadecimal*; *denary*.

base load (*elec*) The underlying steady demand on an electrical power supply system; usually supplied by coal-fired, hydro-electric or atomic power stations.

BASIC Abb. (*comp*) Beginner's all-purpose symbolic instruction code, a high-level computer programming language.

bastard file (*mech*) The cut of any engineer's file that is less coarse than a rough file but more rough than a second cut file. The spacing of the teeth will depend on the length of the file. It is the roughest cut of file in normal use and is used where rapid metal removal is more important than the surface finish.

BA thread (*mech*) See: British Association screw-thread system.

battery (*elec*) Several electric cells joined together. When connected in *parallel*, they have the same voltage as a single cell but a greater capacity and lower internal resistance. When connected in *series* the voltage of each cell is added together, as in a car battery where six cells are connected in series to give a total (nominal) potential of 12 volts. The internal resistance also increases, being the sum of the internal resistance of the individual cells.

baude (*comp*) A measure of the rate of data transfer across the serial interface of a digital communication system in bits per second.

BCD Abb. (*comp*) Binary coded decimal.
beam (*mech*) Any horizontal structural member designed to carry vertical loads. It may consist of an H-section hot-rolled steel joist (RSJ), a hot-rolled British standard beam (BSB), or a fabricated structure. The bending of beams is a major topic of study in mechanical and structural engineering.
bearing (*mech*) A coupling providing support but allowing relative motion between two components. Friction is kept to a minimum by a suitable choice of materials at the interface and by the use of lubrication to keep the two bearing surfaces apart, by use of low-friction materials such as *PTFE*, or by the use of rolling bearings. The majority of machine components are plain journal bearings consisting of a shaft rotating in a lubricated shell with a bearing alloy surface. See: *plain bearing*; *ball-bearing*; *roller bearing*; *air bearing*.
bed plate (*mech*) A cast iron or steel base forming the foundation for the support and alignment of machine elements, e.g. an internal combustion engine coupled to an electric generator.
Belleville spring (*mech*) A diaphragm spring used to apply the friction thrust and also the release action of a diaphragm spring *clutch*. The spring takes the form of a steel disc with a hole through the centre and radial slots. Named after Julian Francois Belleville (nineteenth century French engineer).
Belt drive (*mech*) A friction drive system using pulleys and belts. **1** V-belt drives are used singly or in multiples according to the drive to be transmitted. They are used for short centre drives using V-grooved pulleys and an example is shown in Fig. B.1(a). **2** Flat belt drives are used for longer centre drives using crowned pulleys with or without flanges and may be open or crossed as shown in Fig. B.1(b). Note that only flat belts can be crossed. Both types of belt can be used with compound drives as shown in Fig. B.1(c). See: *toothed belts*; *V-belts*.
Bench drilling machine (*mech*) See: *drilling machine*.

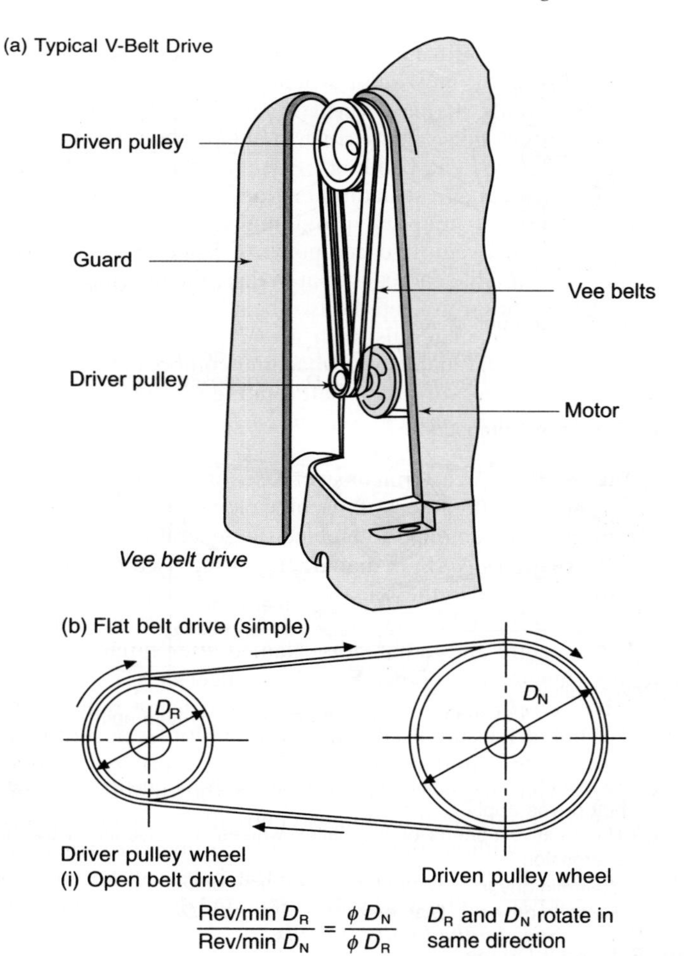

$$\frac{\text{Rev/min } D_R}{\text{Rev/min } D_N} = \frac{\phi D_N}{\phi D_R}$$

D_R and D_N rotate in same direction

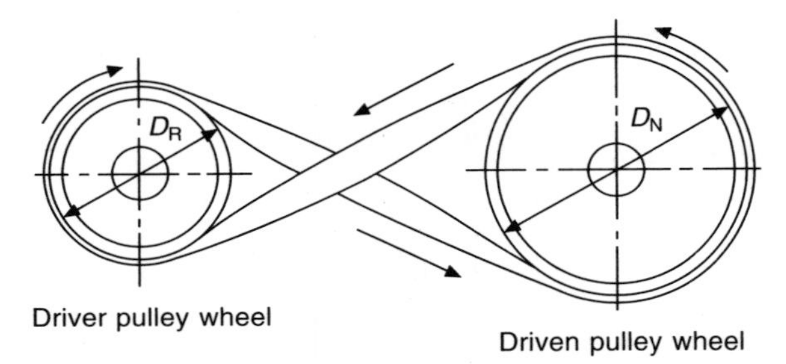

(ii) Crossed belt drive

$$\frac{\text{Rev/min } D_R}{\text{Rev/min } D_N} = \frac{\phi\, D_N}{\phi\, D_R}$$ D_R and D_N rotate in opposite directions

(c) Belt drive (compound)

D_R
D_2
D_1
D_N
Intermediate driven pulley
Driver pulley
Layshaft
Intermediate driver pulley
Driven pulley

(1) To identify the direction of rotation, the rules for open and crossed belt drives apply

(2) The relative speeds of the pulley wheels are calculated by the expression

$$\frac{\text{Rev/min driver}}{\text{Rev/min driven}} = \frac{\text{Diameter } D_1}{\text{Diameter } D_R} \times \frac{\text{Diameter } D_N}{\text{Diameter } D_2}$$

Fig. B.1 Belt Drives

bending equation (*sci*) General equation relating the maximum bending moment, the stress, and the radius of curvature in the bending of beams:

$$\frac{M}{I} = \frac{\sigma}{y} = \frac{E}{R}$$

where M is the maximum bending moment, I is the second moment of area of the beam section about the neutral axis, σ is the stress at the outer fibres of the beam material, y is the distance from the neutral fibres to the outer fibres, E is Young's modulus of elasticity for the beam material, and R is the radius of curvature to the neutral axis.

bending moment diagram (*sci*) A graph showing the bending

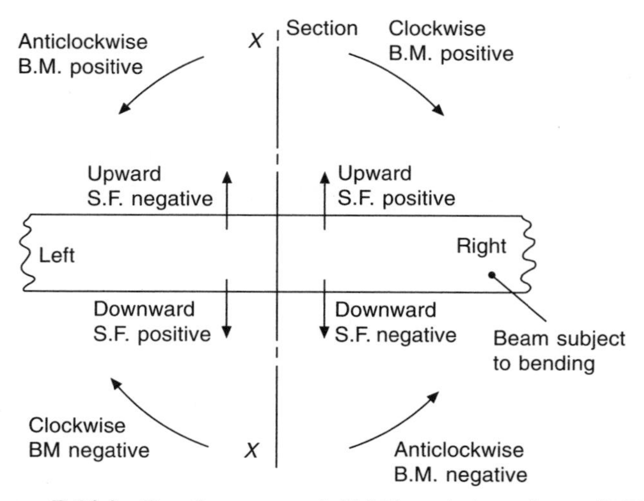

Figure B.2(a) Bending moment (B.M.) and shear force (S.F.) sign convention

moment along the length of a beam, the horizontal (x-axis) representing the length of the beam and the vertical (y-axis) representing the bending moment magnitude as shown in Fig. B.2(a–c).

Bernoulli's law (*sci*) States that when an incompressible fluid flows at a steady rate through a pipe the sum of the kinetic energy, potential energy and the pressure energy of a unit mass of the fluid at any point is constant. See also: *law of conservation of energy*.

Bessemer process (*matl*) An obsolescent method of producing steel by pouring molten iron from the blast furnace into a converter and then burning-out the carbon and impurities by blowing air

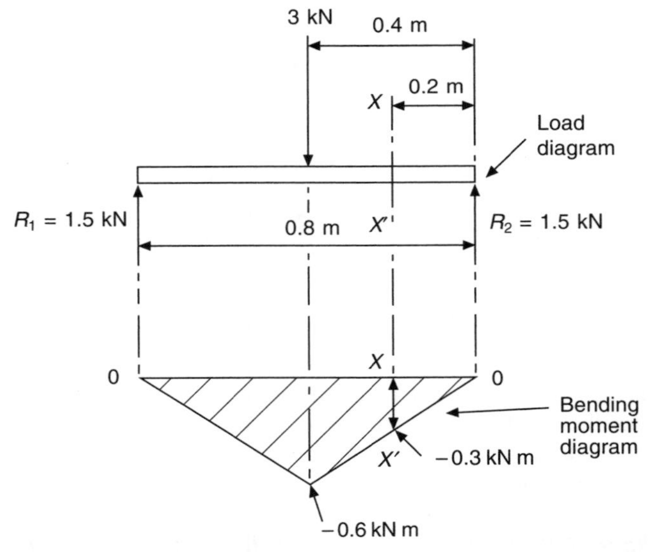

Figure B.2(b) Bending moment diagram for a point load

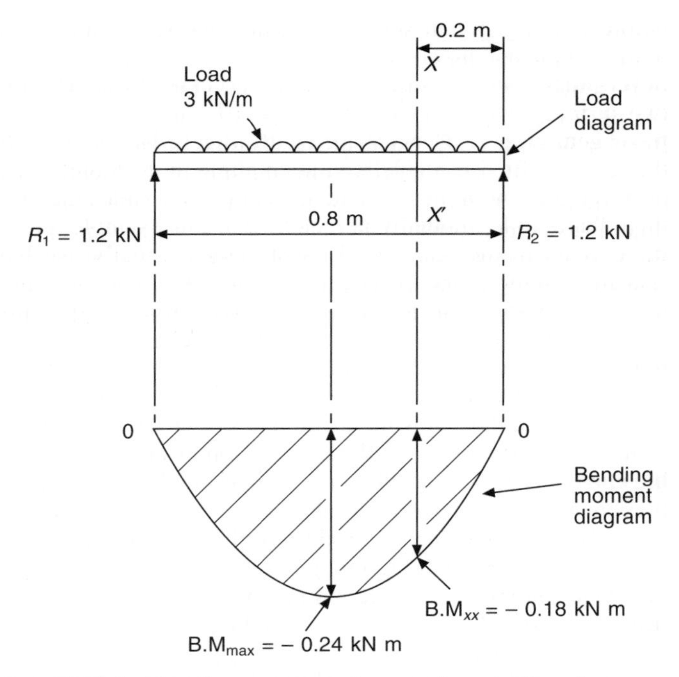

Figure B.2(c) Bending moment diagram for a uniformly distributed load

through the charge. The required amount of carbon and manganese was then added back. The process was widely used when the molten iron had a high phosphorous content (the basic steel making process). The furnace lining was made from firebricks that would combine with the phosphorous and remove it from the steel. If not removed the presence of phosphorus would weaken the steel. The phosphorous-rich slag on the surface of the steel was removed

before pouring. When solid, the basic slag was ground up and used as a phosphorous-rich fertilizer.
bevel gauge (*mech*) Instrument used for checking the angle between two surfaces. The gauge is set to a specific angle before use.
bevel gear (*mech*) Gears shaped with wheel edges at an angle so that gear shafts can be inclined at an angle to each other. A pair of bevel gears with the teeth inclined at (45°) enables the axes of their shafts to be mutually perpendicular. For example, the final drive gears to the rear wheels of a vehicle usually use bevel-shaped meshing gears, the larger the crown wheel and the smaller the bevel pinion gear, to redirect the drive from the gearbox or propeller shaft through 90° to the road wheels.
bias (*elec*) A non-signal current or potential applied to a thermionic valve or a transistor amplifier to ensure operation at the optimal working point, thus allowing maximum symmetrical signal swing of the output waveform without distortion or clipping.
big end (*mech*) Bearing allowing relative angular motion between the connecting rod and the crankshaft of a reciprocating mechanism. For example, the connecting rod transmits the force acting on the piston to the crankshaft big-end journal, converting linear motion into rotary motion in an internal combustion engine.
billet (*mech*) Unfinished steel product which has been hot-worked from a cast ingot.
bimetallic strip (*mech*) Strip comprised of two layers of metal with differing coefficients of thermal expansion. On heating, one layer of metal will expand more than the other, causing the strip to bend. This can then be used to open or close electrical contacts at a desired temperature. See: *thermostat*.
binary (*comp*) Base 2 numbering system which only uses the digits 0 and 1. The binary system is the natural choice for representation of numbers in digital electronics and allows a digital level, or a binary digit, to represent part of a number. See: *bit*, *byte*; *hexadecimal*.

binary coded decimal (BCD) (*comp*) The natural binary system of numbers requires a large number of bits to represent a denary number having several significant figures. Further, the binary number gives no real indication of the size of the denary number. This problem can be overcome by using the *binary coded decimal* (BCD) system where a number in denary notation is converted into binary notation using a pattern (decade) of four bits. The units decade represents denary numbers from 0 to 9, the tens decade represents denary numbers from 10 to 99, the hundreds decade represents denary numbers from 100 to 999, and so on. For example, the denary number 2917 = 0010 1001 0001 0111 in binary.
binary digit (*comp*) See: *bit*.
binomial distribution (*stats*) As its name implies, the binomial distribution deals only with two numbers, i.e. the probability of an event occurring (p) and the probability of an event not occurring (q). Therefore, if p is the probability of an event occurring and q is the probability of an event not occurring, then the probabilities that the event will occur 0, 1, 2, 3, 4, … n times in n trials are given by the successive terms of the expansion of $(q + p)^n$ taken from left to right. See: *binomial theorem*.
binomial theorem (*maths*) A binomial expression is one that contains two terms, e.g. $(a \pm b)$, $(c \pm d)^2$, $(2m \pm n)^3$. The binomial theorem is a formula for raising a binomial expression to any power without lengthy multiplication. The general binomial expansion of $(a + b)^n$ is given by:

$$(a + b)^n = a^n + na^{n-1}b + \frac{n(n-1)}{2!}a^{n-2}b^2 + \frac{n(n-1)(n-2)}{3!}a^{n-3}b^3 + \ldots + b^n$$

BIOS Abb. (*comp*) Basic input–output system. The basic level of the *operating system* of a computer, stored in the ROM, which communicates directly with the hardware; higher level functions of the operating system are based on the BIOS.

bipolar (junction) transistor (BJT) (*comp*) Common type of transistor which has three junctions: a base, a collector and an emitter. The collector current is controlled by the base current and the transistor behaves as a current amplifier.

bistable circuit (*elec*) A circuit that has two stable states. It will remain in a given state until it is triggered, when it will change state and remain in the new state until it is triggered again. A bistable circuit has two inputs and two outputs, as shown in Fig. B.3 together with its truth table. When $\overline{Q}$ = 1 then Q = 0 and

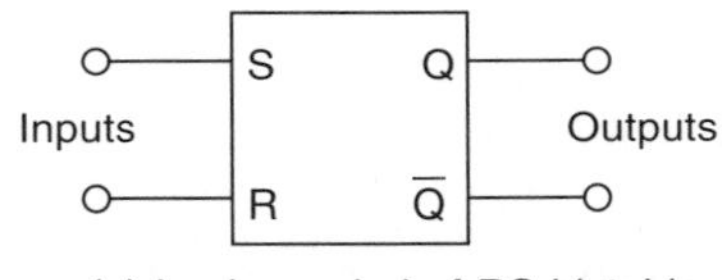

(a) Logic symbol of RS bistable

S	R	Q	Q+1	
0	0	0	0	No pulse on S or R; Q stays as it was.
0	0	1	1	No pulse on S or R; Q stays as it was.
1	0	0	1	Pulse on S sets Q to 1.
1	0	1	1	Since Q = 1 pulse on S = no change.
0	1	0	0	Since Q = 0, pulse on R = no change.
0	1	1	0	Pulse on R resets Q to 0.
1	1	0	X	Indeteminate
1	1	1	X	Indeteminate

(b) RS bistable truth table

Fig. B.3 The Bistable Circuit

when $\overline{Q}$ = 0 then Q =1. From the truth table it can be seen that the state after the application of a set or reset pulse is Q + 1. The indeterminate state would occur when S = R = 1 and Q could be 0 or 1. This condition is not allowed. See: *monostable circuit.*

bit (*comp*) a single binary digit, i.e. the individual 1s or 0s in base 2. See: *binary*; *byte.*

black body (*sci*) A term for a theoretical body that absorbs all thermal *radiation* falling on it and emits the maximum amount of radiation possible; this has nothing to do with colour. In reality all bodies emit or absorb thermal radiation at a lower level than a black body but it is a useful concept for study.

blackheart process (*matls*) Process for malleablizing 'white' cast iron, more common in America than the UK. The castings are packed in airtight boxes out of contact with atmospheric oxygen and heated at 850 to 950°C for 50 to 170 hours, depending on the mass and thickness of the castings. The effect of this prolonged heating is to cause the iron carbide (cementite) to break down into small rosettes of graphite. The final structure is comprised of crystals of ferrite and fine graphite particles. The fine graphite particles darken the colour of a fractured surface of cast iron after treatment and it is this darkening that gives the process its name. See: *whiteheart process.*

blast furnace (*matls*) A furnace for smelting iron ore in order to extract crude metallic iron. The charge is in direct contact with the burning fuel and products of combustion. Typically the structure is 60 m high, 7 m diameter at the base and operates continuously. The furnace is fed with ore, coke and limestone and gets its name from the hot air blast of combustion air blown at its base. At intervals the slag and molten iron are run off. The coke burns to produce carbon monoxide and this reacts with the ore to produce iron and carbon dioxide. The limestone combines with some of the impurities in the original mineral ore to form a molten slag

that floats on the surface of the liquid iron and is tapped off before the iron.
bleeding (*sci*) **1** The removal of trapped air from liquid systems, such as hot water heating or power hydraulics. **2** A term used in steam power plants to approximate a regenerative cycle where a small amount of steam is taken from the high pressure turbine and used to heat the feed water.
block (*mech*) The casing of the pulley wheels of a lifting tackle that supports the axle and provides an attachment point.
block diagram (*elec*) Method of representing the individual functional elements of a process or a system by boxes interconnected by arrows representing the signal paths between the elements. Each box represents a relationship between its inputs and outputs. The usual convention is to show the overall system input at the left side of a page and the output at the right, as shown in Fig. B.4.
bloom (*mech*) Semi-finished metal product of rectangular cross-section (the first stage of hot-rolling a cast ingot), it is larger than a *billet*.

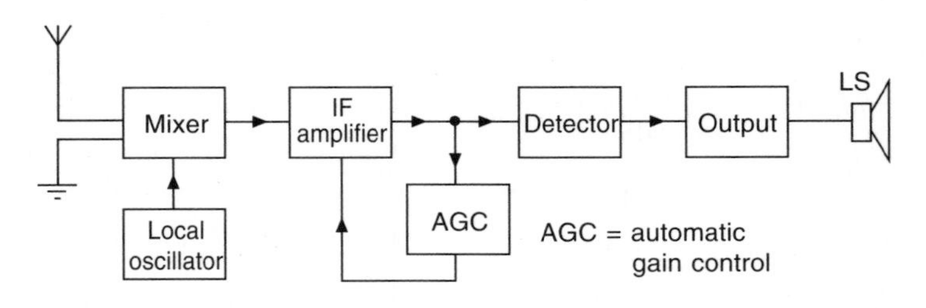

Fig. B.4 Block Diagram for a Radio Receiver

blowdown (*mech*) Process of releasing water from a steam boiler and replacing by fresh feed water in order to dilute the impurities. Water that enters a boiler contains impurities and the steam produced is virtually pure; therefore the concentration of impurities in a boiler will gradually increase, causing operational problems. Blowdown is carried out to keep the impurities down to an acceptable level. The turbulence caused by blowdown also helps to dislodge and remove any build-up of limescale on the boiler tunes and plates.

blown casting (*mech*) A casting containing bubbles caused by gases or steam in the molten metal. Large bubbles just below the surface are called blowholes, and smaller bubbles dispersed through the casting result in porosity.

blowpipe (*mech*) Also called a *blow torch.* A device in which flammable gases are mixed to be burnt at the nozzle on the end of the blowpipe. Natural gas and pressurized air, or bottled gas and air may be used for the moderate temperatures required for soldering and brazing, bottled gas and oxygen for braze welding, and acetylene and oxygen for fusion welding. The blowpipe has separate vales for the fuel gas and air or oxygen so that the mixture can be controlled to suit the job in hand. The gas is supplied via flexible hoses.

blueprint (*mech*) An obsolete method of reproducing an engineering drawing or plan, in which the tracing was placed in contact with a sensitised paper and subjected to intense ultraviolet light. The process got its name from the fact that the drawing was reproduced in white on a blue background. Although superseded by the quicker and more efficient *dyeline* process (black lines on a white background) after the Second World War, drawings reproduced from a tracing were still popularly referred to as blueprints. A term still used in the media for any plan or scheme. Nowadays computer-aided design systems have removed the need for tracing and hard copy can be printed out quickly as and when required either in black and white or in colour.

Bode diagram (*elec*) A graphical representation in which the gain and/or phase shift of a circuit is plotted against the frequency of the applied signal.

BODMAS (*maths*) An acronym standing for the order of basic arithmetic operations: brackets, of, divide, multiply, add and subtract. This is the order in which all calculations should be approached. Note that in this context 'of' can also mean 'multiply'.
boiler (*mech*) Pressure vessel in which water (or other fluid) is heated and released as steam or hot water. Industrial boilers usually consist of a heat-exchanger, transferring heat from combustion gases of a furnace to water in a pressurized boiler drum, and producing steam for electricity generation, manufacturing processes, heating, etc.
bolster (*mech*) **1** Device for supporting metal around a rivet hole whilst punching the hole to give a cleaner cut. **2** The base plate of a press tool.
bolt (*mech*) Threaded rod used with a nut for fastening two components together. The bolt is held in place by a shaped head (usually hexagonal) whilst a tightening torque is applied to the nut.
bonderizing (*matls*) See: *conversion coatings*.
bonding 1 (*mech*) The process of joining two surfaces by use of an adhesive. **2** (*elec*) The interconnection of metallic components to earth for protection against electric shock.
bonding clip (*elec*) Device for clamping the external metallic components of a system to the earth continuity conductor for earth bonding purposes.
Boolean algebra (*maths*) Form of mathematical algebra that expresses the behaviour of systems that only have two states; it was developed by George Boole in the middle of the nineteenth century. Symbols are used to represent variable quantities as in conventional algebra, the variables using one of two states called true or false, and equations may be manipulated and equated accordingly. It is used extensively by the designers of digital circuits to calculate the logic functions required. Standard arithmetic operations are not used, only the logical operations of AND, OR and NOT.

boot disk (*elec*) Floppy disk that can be used to start up a computer, when it is switched on, if the operating system has become corrupted on the hard disk.
boring (*mech*) Process of enlarging a cast or drilled cylindrical hole in metal by use of a single-point cutting tool mounted in a machine tool such as a lathe (work rotates) or boring machine (cutter rotates). The purpose of boring is to provide a hole with a good positional and dimensional accuracy and a good surface finish.
boss (*mech*) A cylindrical protuberance on a casting or forging. It provides additional strength around a bolt-hole or a hole that supports a shaft. It also allows the surface to be locally machined to produce a flat seating onto which the nut or bolt head can be tightened.
boundary (*sci*) Imaginary line that separates a system from its surroundings for the purposes of investigation or problem solving. The boundary may be fixed or elastic but must be carefully defined.
boundary lubrication (*mech*) a state of partial lubrication (such as occurs in the cylinders of a car when starting from cold) that may exist between two surfaces in the absence of the oil flow necessary for fluid film lubrication. Boundary lubrication results from the adsorption of a mono-molecular layer of lubricant left adhering to the bearing surfaces after fluid film lubrication has ceased layer of lubricant left adhering to the bearing surfaces after fluid film lubrication has ceased.
Bourdon tube (*sci*) Tube used as the sensing element of a pressure gauge, commonly in the shape of a 'C' although other shapes are sometime used such as a spiral. One end is open and applied to the source of fluid whose pressure is to be measured and the other is closed. As the fluid pressure increases the tube tends to straighten. The closed end is connected to a pointer and dial indicator via a gear linkage. In this manner the pressure

signal is converted to a mechanical one that can be applied to a visual indicator as shown in Fig. B.5.

Bow's notation (*sci*) Method of labelling forces in space diagrams and force diagrams. In the *space diagram* (Fig. B.6(a)), the spaces between the forces are labelled with *capital* letters. Each force may be referred to by the two letters either side of the force. On the *force diagram* (Fig. B.6(b)) the corresponding *lower case* letters are then used to label the ends of the vectors as shown. The labelling around the diagrams can be clockwise or anticlockwise, but the same convention must be used in both diagrams to avoid confusion.

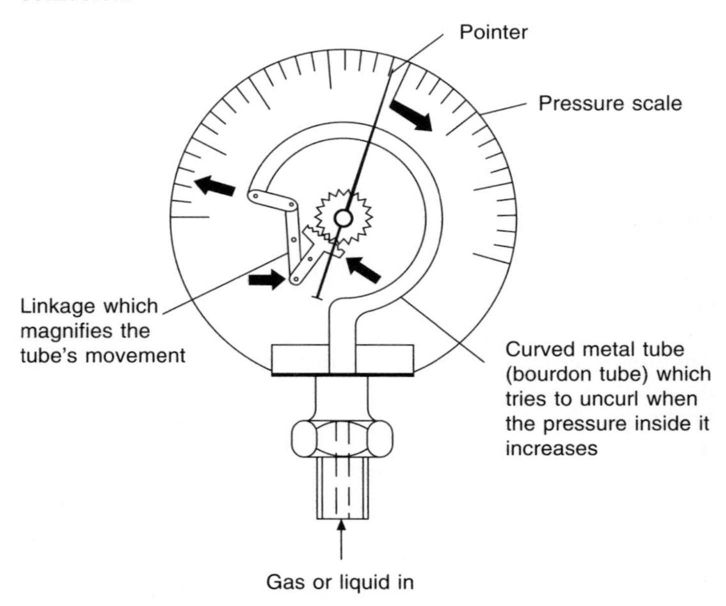

Fig. B.5 Bourdon Tube Pressure Gauge

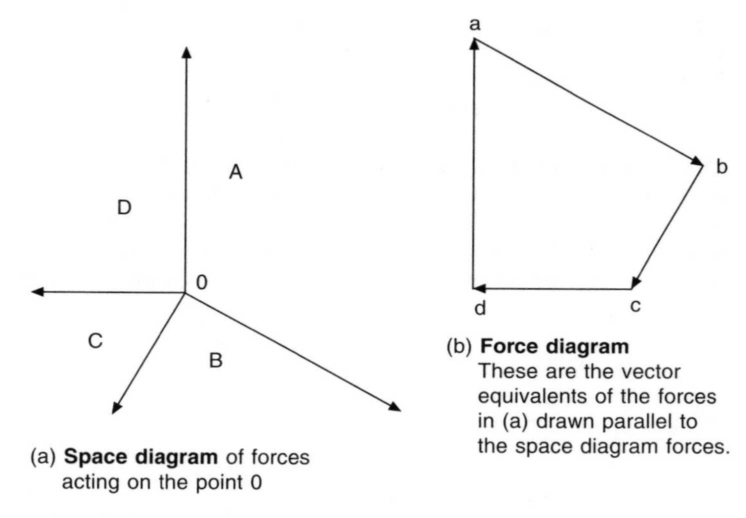

Fig. B.6 Bow's Notation

Boyle's law (*sci*) A law that applies to perfect gases, and which states that for a mass of gas held at a constant temperature the pressure is inversely proportional to the volume. The relationship was discovered by Sir Robert Boyle in the mid-seventeenth century. See also: *gas laws*.

brake (*mech*) Device for retarding or stopping a body in linear or angular motion. Usually this refers to either a rotating machine wheel or a vehicle in motion, where brake pads or shoes of friction material are forced onto a rotating disc or drum and the frictional forces allow the kinetic energy to be converted into heat energy.

brake power (*sci*) Standardized test procedure for measuring the power generated by the crankshaft of an engine, so called because the engine torque may be measured using a brake to resist the rotation of the flywheel and the brake power measured as follows:

$$\text{brake power (kW)} = \frac{2\pi\, TN}{60\,000}$$

where T is the measured torque in Nm and N is the engine speed in rev/min. The brake power is always less than the *indicated power*.

brass (*matls*) An alloy of copper and zinc sometimes containing small amounts of other metals such as tin, aluminium, iron and lead. The composition of brass alloys varies widely depending on the intended use; for example, naval brass (62% copper, 37% zinc, 1% tin has good corrosion resistance; common brass (63% copper, 37% zinc) is a good general-purpose brass; cartridge brass (70% copper, 30% zinc) has the maximum ductility possible in any brass alloy.

braze welding (*mech*) Also called *bronze welding* is a process that differs fundamentally from brazing or hard soldering. The filler material used is a bronze that is much stronger than brazing spelter. The molten filler material is less fluid than spelter and is not drawn into the joint by capillary action. Instead, it is deposited so as to form a fillet or 'bead' as in fusion welding. Like soldering and brazing, the joint surfaces remain unmolten throughout the process, the molten filler material forming an amalgam with the parent metal at the joint surfaces. The temperatures involved are higher than for brazing and an oxy-propane torch is normally used. A proprietary flux has to be used and this is supplied by the manufacturer of the bronze filler material.

brazing (*mech*) A method of joining two pieces of metal together by causing a molten brass alloy (*spelter*) flow into the joint by capillary attraction. The spelter is not just keyed to the surfaces mechanically but forms an amalgam with the joint faces which do not themselves become molten. This results in a very strong joint. The process gets its name from the brass alloy used for the spelter. The joint so formed is stronger than soldering. The metal pieces to be joined by brazing have to be clean both mechanically and chemically (free from surface oxides). A borax-based flux is used. The parts to be joined are fitted together and heated to red heat by a gasblow torch. Spelter and a flux are applied to the junction. The spelter melts and is drawn into the slight gap between surfaces, forming a solid brazed joint upon solidifying. Although strictly incorrect, the term 'brazing' is frequently used for silver soldering (hard soldering).

breadboard (*elec*) Circuit board for constructing temporary trial versions of electronic circuits for testing prior to permanent construction. Consists of a plastic board with rows of holes spaced to receive integrated circuits or other components and wiring, and conductors to other holes for connection to a power supply. Other types of breadboard are available for constructing permanent circuits and consist of a card with holes to secure components and wiring for soldering. Copper strip conductors run along the board for ease of construction and connection to power supplies.

breakdown maintenance (*mech*) Maintenance programme (or rather the lack of one) that involves waiting for a system to break down before any maintenance work is carried out on it. Only effective when the cost of breakdowns is less than that of preventative maintenance. See also: *planned maintenance*.

bridge gauge (*sci*) Measuring device for monitoring the relative movement between two parts of a mechanism resulting from the wear of components in that mechanism.

Brinell hardness (*matls*) A test for measuring the hardness of materials resulting in a Brinell hardness number which has no units. A hardened-steel ball of diameter D is forced into the surface of the test piece of material by a specified load F. The diameter of the indentation (d) is measured by a specially calibrated low-powered microscope. The Brinell hardness number can then be calculated by the expression:

$$H = \frac{F}{\text{surface area of indentation}}$$

or

$$H = \frac{F}{\frac{\pi D}{2}\left[D - \sqrt{(D^2 - d^2)}\right]}$$

In practice, the hardness number is generally found in a series of tables for various values of *F*, *D* and *d*. Note that the relationship between the load and the diameter of the hardened-steel ball indenter is given by the expression $F/D^2 = K$, where for ferrous metals $K = 30$; for copper and copper alloys $K = 10$; for aluminium and aluminium alloys $K = 5$; and for lead, tin and white bearing metals $K = 1$. Owing to the fact that the hardened-steel ball indenter tends to collapse and flatten when testing hard materials, this test has largely fallen out of use in favour of the Vicker's diamond pyramid test. See: *Vicker's hardness test.*

British Association thread (*mech*) A metric thread developed by the British Association for use in scientific instruments. The thread has a V-form with an included angle of 47.5°, and well-rounded crests and roots as shown in Fig. B.7. The thread sizes

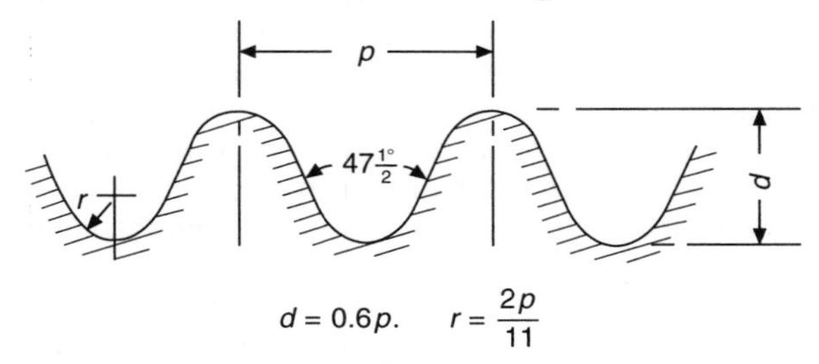

Fig. B.7 British Association (BA) Threads

range from 0 BA (maximum diameter 6 mm × pitch 1 mm) to 10 BA (maximum diameter 1.7 mm × pitch 0.35 mm). Now largely superseded by the miniature metric series.

British standard brass (BSB) (*mech*) A constant pitch screw thread system with a British Standard Whitworth profile suitable for thin-walled tubing. By keeping the pitch independent of the thread diameter, the depth of the thread does not break through the wall of the tube whatever the diameter.

British standard fine (BSF) (*mech*) A Screw thread with a British Standard Whitworth profile but with a finer pitch for a specified diameter, used for screws subject to vibration.

British standard pipe (BSP) (*mech*) A constant-pitch screw thread system with a British Standard Whitworth profile sized by the bore of the pipe on which it is cut. It can be parallel or tapered.

British Standards Institution (BSI) (*mech*) A national establishment for the preparation and publication of approved specifications and codes of practice to ensure the control of interchangeability, performance, conformance and quality of standard materials and components.

British Standard Whitworth (BSW) (*mech*) The first standard screw thread system to be devised. It was devised by Sir Joseph Whitworth (a nineteenth century British Engineer). It had inch dimensions and the V-form with its 55° included angle is shown in Fig. B.8. Many other screw thread systems are based on this thread form. Although now obsolescent and no longer recommended for new designs it is still widely used for maintenance purposes. Also the ISO metric pipe thread is based on the Whitworth form. More modern thread sytems both in metric and inch dimensions favour the 60° included angle. (See: *unified threads*; *ISO metric threads*).

British thermal unit (*sci*) An obsolescent unit of heat energy, symbol Btu, defined as the heat energy required to raise the temperature of a 1 lb mass of water by 1°F.

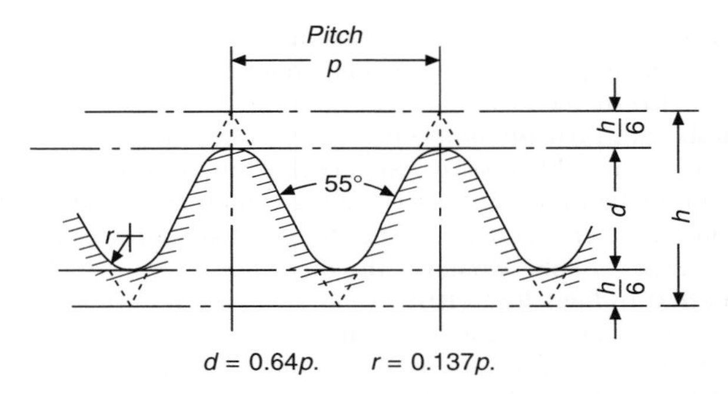

Fig. B.8 Whitworth and British Standard Fine Threads

brittleness (*mech*) Material property that is the opposite of toughness, i.e. a material that will fracture under shock loads, or is able to fracture without deformation.

bronze (*matls*) A family of alloys based on the metal copper. The most common are copper and tin plus a deoxidizer, which is zinc in gun metal and phosphorus in phosphor bronze. Admiralty gunmetal contains 88% copper, 10% tin and 2% zinc. It was originally used for naval cannons but is now used for marine pump and valve bodies as it casts well and has good corrosion resistance. Cast phosphor bronze, used for high-duty bearings, contains 10–18% tin and up to 0.25% phosphorus, the remainder being copper. Leaded bronze containing 5% tin and up to 20% lead is used as a bearing alloy where a less-rigid metal is required. Wrought, low-tin bronzes containing up to 4% tin and up to 0.25% phosphorus can be rolled into sheet or drawn into spring grade wire, or used for electrical contact blades. Drawn phosphor bronze containing up to 6% tin and up to 0.25% phosphorus is used in the work-hardened condition for turned components requiring high strength and corrosion resistance such as steam valve spindles. In addition to the tin bronzes, aluminium bronze alloys are used where corrosion resistance at high temperatures is required in steam and chemical plants. They are more expensive than the tin bronzes. See: *aluminium bronze*; *cupro-nickel alloys*.

brush (*elec*) A carbon block that presses against the commutator segments or the slip rings of a rotating electrical machine in order to conduct an electrical current to or from the machine. Carbon is used because of its inherent lubricating properties, which reduce wear, and because its relatively high resistance, compared with metals, reduces arcing at the contact face.

brush gear (*elec*) General term used for the brushes and associated components of electrical machines.

bubblejet printer (*comp*) A printer which operates by heating the ink in the printhead to form bubbles. These expand and eject the ink through a nozzle onto the paper.

buffer 1 (*elec*) An electronic amplifier, frequently of unity gain, used for isolating the input circuit from the output circuit. It usually has a high input impedance to avoid loading the driver stage and low output impedance to provide current drive into a transmission line or peripheral device. **2** (*comp*) Computer memory to aid transmission of data between CPU and peripheral device which operates at a different speed.

bug (*comp*) A software fault that prevents the correct operation of a computer program.

bulk modulus (*matls*) The ratio of the direct stress applied to a body, to the fractional decrease in volume.

burn-out (*matls*) A condition arising when a liquid is heated by a submerged surface and the bubbles forming at the surface are so large and numerous that a blanket of vapour forms over the heating surface. As the liquid is prevented from flowing back onto the surface there is a sudden rise in temperature of the surface. Under these conditions the heating surface will be weakened and is often in danger of melting.

bus 1 (*comp*) An arrangement of parallel wires along which signals pass from one component of a computer to another, particularly the *central processing unit (CPU)*, the memory and the input/output section. Typically, the bus is comprised of an address bus, a control bus and a data bus. The address bus is used to send the address of memory locations that are to be written to or read from, or the port address when reading to or from a port. The CPU can address up to 2^n memory locations where n is the number of address lines. The control bus is used to control and synchronize the different components of the computer, typically by sending control signals that indicate whether data is to be sent from the memory or port to the CPU, or from the CPU to the memory or port. The data bus is bidirectional and is used to read or send data from or to memory locations or ports. **2** (*elec*) An electrical conductor used for distributing or collecting electrical energy or signals. A busbar is commonly a conductor of rigid copper construction used in electrical power transmission systems. A number of subcircuits/appliances may be connected to common supply busbars via suitable protective devices such as fuses.
bush (*mech*) A cylindrical component made from a bearing material and fitted into a machine part to support a shaft. Also a hardened cylindrical steel component used to guide a drill in a drilling jig used for repetition machining.
buttress thread (*mech*) Type of thread that provides a force in one direction only. The thread section is a blunted triangle with an included angle of 45°, as shown in Fig. B.9.
butt weld (*mech*) A pressure-welded joint in which the two flat surfaces meet normally (at 90°) to the direction of the other surfaces, e.g. the butt welding of a carbon steel shank to a high-speed steel blank for making large twist drills. Electrical resistance heating is normally employed with the current passing through the two components and the heat being generated at the relatively high

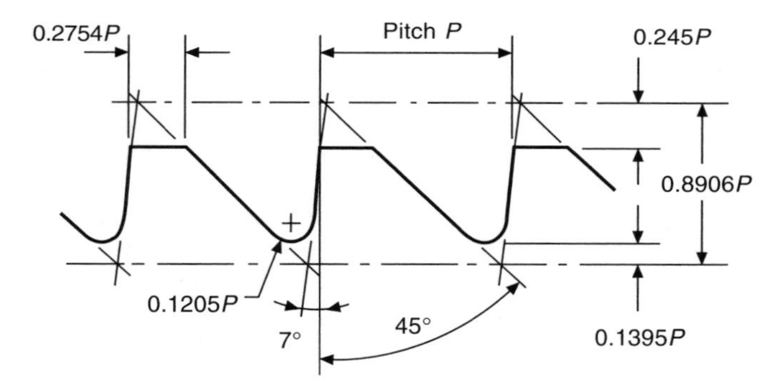

Fig. B.9 Buttress Thread Form

resistance of the contact faces. The welding pressure is applied axially. The principle of this process is shown in Fig. B.10.
byte (*comp*) The size of a binary number consisting of 8 binary digits. See also: *bit*; *nibble*; *word*.

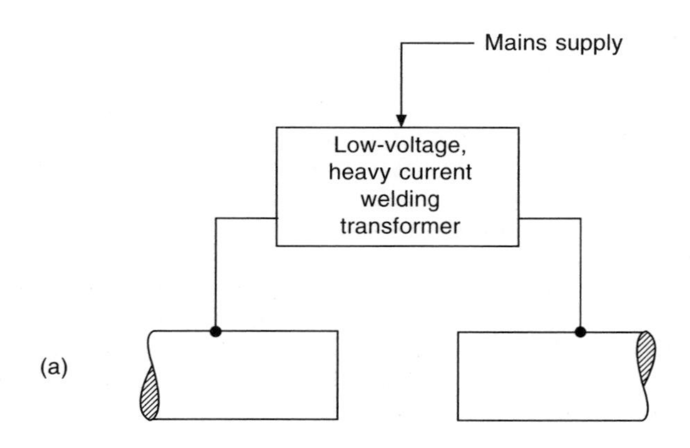

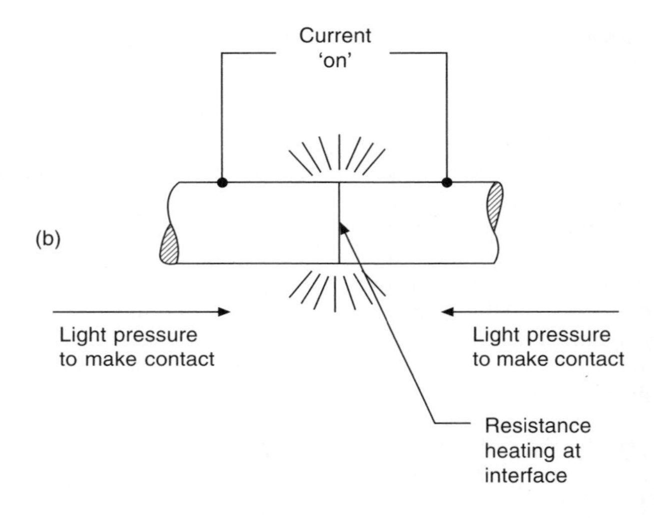

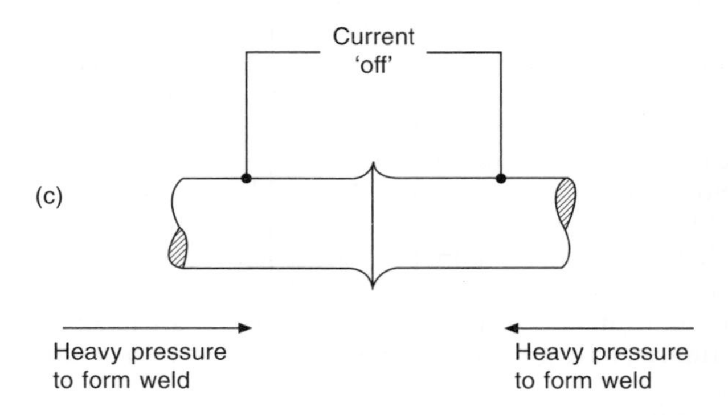

Fig. B.10 Butt-weld (Resistance)

C

C (*comp*) High-level computer programming language, in common use in the engineering industry. It is the language in which the UNIX operating system is written.
C++ (*comp*) An enhancement of the C programming language to include object-oriented programming.
cache (*comp*) High-speed built-in memory of computer memory management unit. Descriptors of currently used segments are kept in the cache memory for faster access than if they were in the main memory.
CAD (*CAE*) See: *computer-aided design.*
CAD/CAM (*CAE*) The merging of the digitized data produced during the design process with the data required for manufacture. Industry-standard CAD packages can export DXF (drawing exchange files) in a format that can be imported by many CAM packages including the IGES (initial graphical exchange system). Computer aided manufacturing systems such as PEPS (production engineering productivity system) can import DXF files. Once the program has been proved by simulation, then the simulation can be post-processed into any CNC language and downloaded directly into the production machines. See: *computer-aided manufacture.*
CAL Abb. (*comp*) Computer-assisted learning.
calculator (*maths*) Portable, pocket-size digital computer, dedicated to performing mathematical functions. Modern ones are becoming more sophisticated and may be programmable to solve formulae, have graphical displays and able to work with different bases.
calculus (*maths*) A set of techniques for calculating quantities

that vary rather than being fixed. See: *differential calculus*; *integral calculus*.

caliper (*mech*) An instrument for transferring the distance between two parallel surfaces for measurement against a rule. They can be used for external or internal dimensions and of either the firm-joint or spring-joint type, the latter being controlled by a knurled nut. See: *micrometer caliper*; *vernier caliper*.

calorie (*sci*) The centimetre–gram–second (cgs) unit of heat defined as the quantity of heat energy required to raise 1 gram of water by 1°C. Now largely replace by the SI unit of heat energy, the *joule*(J); 1 calorie = 4.186 joules.

calorific value (*sci*) A measure of the amount of heat energy that any fuel can provide. It is measured as energy per unit of mass, usually MJ/kg.

calorifier (*sci*) Tank for heating water, the heat being supplied by a submerged coil of heated piping.

calorizing (*matls*) Process of spraying the surface of steel with aluminium followed by heating to between 800°C and 1000°C, in order to provide resistance against oxidation corrosion. See: *cementation processes*.

cam (*mech*) A rotating lobe-shaped component, with a predetermined contour, that imparts linear motion to a mating component called the cam follower which bears upon the contour of the cam. See Fig. C.1.

cam follower (tappet) (*mech*) Any component that is in contact with the contoured (lobed) surface of a cam, and which converts the rotary motion (angular displacement) of the *cam* into cyclical reciprocating motion.

camshaft (*mech*) A machine shaft carrying several *cams* that can convert rotary motion (angular displacement) of the shaft into the complex linear motion of cam followers. In an *automatic lathe* the camshaft carries the cams that actuate such functions as opening and closing the collet chuck, traversing the tool turret

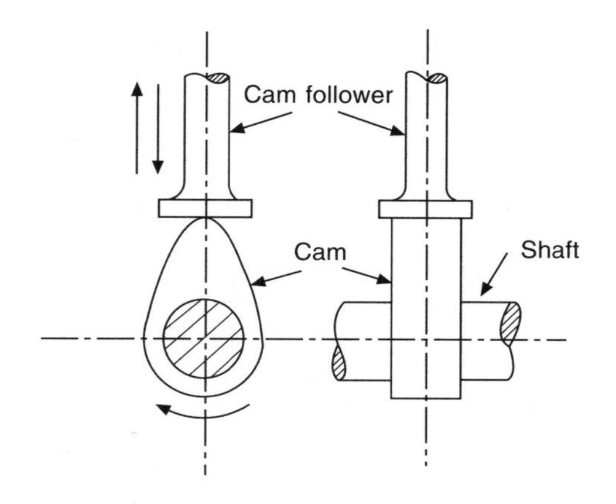

Fig. C.1

and providing in-feed to the cross-slide tools in the correct sequence.
canned cycles (*CAE*) These are standard fixed cycles that save the need for repetitive programming of common operations. The sequence of events for a standard cycle is embedded in the memory of the controller's computer at the time of manufacture and is called up, when required, by an appropriate G-code. For example, G81 calls up the drilling cycle where the sequence of events, as shown in Fig. C.2, is: rapid traverse to first hole, rapid traverse to clearance plane height, feed to depth of hole, rapid traverse out to clearance plane height, rapid traverse to next hole, where this is repeated for as many holes as required.
cantilever (*sci*) A beam that is rigidly supported at one end only, e.g. by building one end into a solid abutment.
capacitance (*elec*) The ratio of the charge on a conductor or system of conductors to the potential difference across the conductor or system. This can be expressed as $C = Q/V$, where C is the

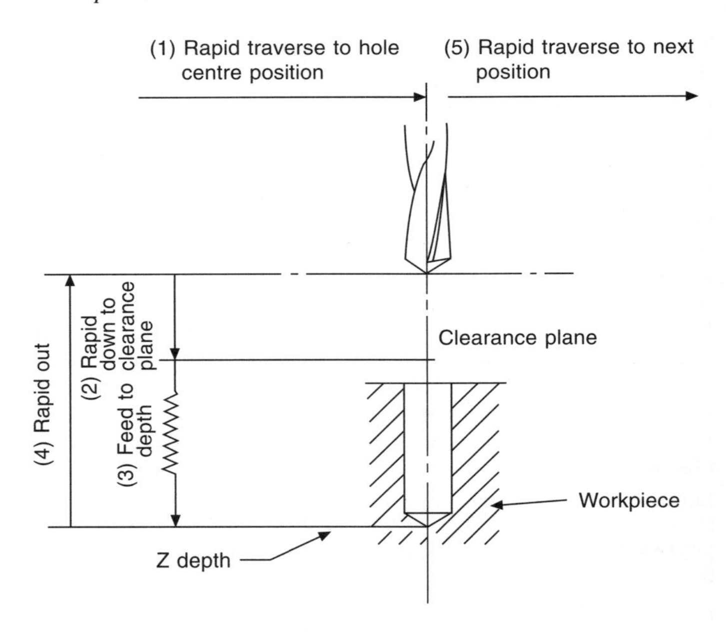

Fig. C.2 Canned Drill Cycle (G81)

capacitance measured in farads, Q is the quantity of electrical charge measured in coulombs and V is the potential difference across the system measured in volts. One farad of capacitance of a system requires a charge of one coulomb to raise the potential by one volt. This is a very large charge, and typical practical values are measured μF, nF and pF.

capacitor (*elec*) An electronic component consisting of conductors (plates) separated by insulating material (dieletric) and designed to have a specific *capacitance*. A capacitor can store an electrical charge in *d.c.* circuits, or create a reactance within *a.c.* circuits.

The greater the area of the conductors, and the thinner the dielectric, the greater will be the capacitance. The material from which the dielectric is made will also affect the capacitance.

capacitor start (*elec*) Method of starting electric single-phase motors by using series capacitance to advance the phase angle of the current.

capstan lathe (*mech*) A development of the centre lathe used for repetition machining when operated by semi-skilled labour. The conventional tail stock (see: *centre lathe*) is replaced by a turret (capstan) mounted on a longitudinally moving slide in a saddle fixed to the bed. Preset tooling is mounted on the turret so that each tool can be presented to the work sequentially for successive turning operations. After each operation, the capstan automatically revolves (indexes) bringing the next tool into position, saving time on machining operations. There is also a cross-slide for such operations as thread chasing and parting-off. The use of capstan lathes has largely died out with the advent of the CNC turning centre.

carbides (*matls*) Compounds of carbon with a metal such as tungsten and/or titanium. Such carbides are very hard and are used, for example, for cutting-tool tips.

carbon fibre (*matls*) Fibre used to strengthen composites in order to produce materials that have a high strength-to-weight ratio. Produced by heating a suitable fibre under tension to a temperature where the fibre decomposes leaving behind a long chain structure of carbon atoms. The fibre is then moulded in a matrix of thermosetting resin to produce a rigid, high-strength and light-weight material. Carbon fibre is now used for sports equipment such as racket frames and skis, by some sports car manufacturers, and in high technology military aircraft.

carburation (*sci*) The mixing of air with the correct proportion of vaporous fuel, usually petrol, to form a fuel mixture for combustion in an internal combustion engine. Carburation is performed by a device called a carburettor in which the liquid fuel is introduced into a stream of air to produce small droplets. See also: *fuel injection.*

Carnot cycle (*sci*) An ideal reversible heat-engine thermodynamic cycle of operations for maximum thermal efficiency. One cycle consists of an *isothermal expansion*, an *adiabatic expansion*, an *isothermal compression* and an *adiabatic compression.*

carriage (*mech*) A flat casting that slides along on the bedways (shears) of a centre lathe (see: *centre lathe*), and which has guides on the top surface for the *cross-slide* perpendicular to the grooves of the bedways. The *apron* is attached to the front of the carriage and carries the controls for the carriage movements. The combined assembly of the carriage and apron is referred to as the *saddle.*

carrier (*mech*) Device used to rotate the workpiece on a lathe when turning between centres and where no chuck is used. The carrier has a hole through it to take the bar to be turned and a clamp screw to tighten it onto the bar. A projecting dog on the catch-plate which is mounted on and rotated by the lathe spindle engages with the carrier to provide the turning force on the work.

Cartesian coordinate robot (*CAE*) These have three orthogonal sliding axes as shown in Fig. C.3 and they have the same positional resolution, accuracy and repeatability as for a CNC machine tool. It, alone amongst robot configurations, has its spatial resolution constant in all axes of motion and throughout its work envelope (volume).

case hardening (*matls*) The hardening of the surface of tough, low-carbon steel components by altering the composition of the outer layer of steel prior to heat treatment. This is achieved by one of the following three carburizing techniques. (1) Heating the component to be treated in a sealed and airtight box with a solid carbon-rich material to a temperature of 900°C to 950°C for several hours depending on the depth of the case required. This is the

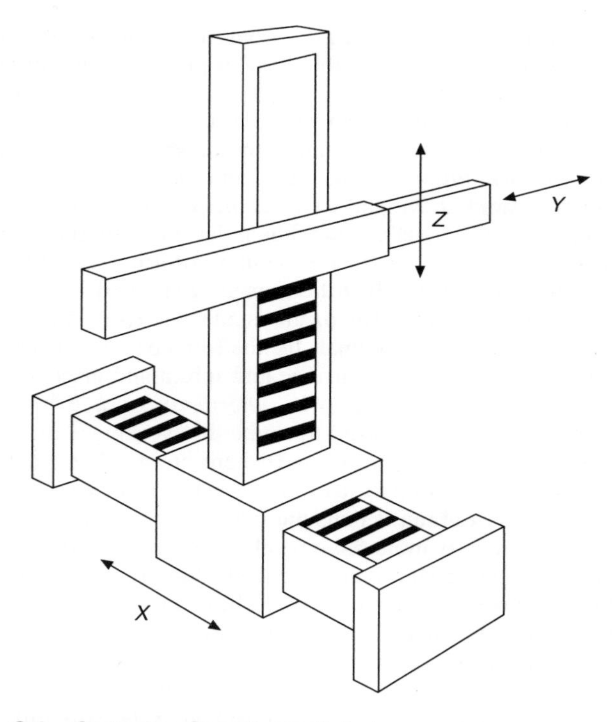

Fig. C.3 Cartesian Coordinate Robot

most common method used in workshops. (2) Immersing the component in a bath of molten salt consisting of sodium cyanide and sodium carbonate, causing absorption of carbon. (3) Heating the component to the required temperature for several hours in a furnace containing a gaseous atmosphere, usually a hydrocarbon. All these processes have the same aim, i.e. to increase the carbon content of the outer layer of the steel so that it will be susceptible to quench hardening. After carburizing, the components are raised to the correct temperature for hardening a high-carbon steel and quenched (cooled rapidly) in water or oil. This process produces components with a hard case and a tough, shock-resistant core.

casting (*mech*) Process of shaping metal by pouring the molten metal into a mould having a cavity of the required shape, prior to machining once the metal has solidified. The most common method of casting is sand casting where the mould form is made in sand in a pair of mating boxes by a wooden pattern, similar in shape to the required component. The metal is then poured in and allowed to cool. In die-casting the mould shape is formed in metal. Castings are made in steel, cast iron, brass, aluminium and in many other metals and alloys. See: *pattern*; *patternmaker*; *shell mouding*; *lost wax process*; *investment casting*.

cast iron (*matls*) A ferrous metal consisting of iron together with 2.0–4.5% carbon, up to 0.65% manganese, up to 2.75% silicon, and traces of sulphur and phosphorus as residual impurities. Common, grey cast irons have a relatively low tensile strength (UTS between 150 MPa and 400 MPa) compared with steel and tend to be brittle. Components made from cast iron should only carry non-impact compressive loads. Compared with their relatively low UTS, cast irons have compressive strengths ranging between 600 MPa and 1200 MPa. The high carbon content does, however, make machining easy and it is easy to cast. See: *malleable cast irons*; *spheroidal graphite (SG) cast irons*; *alloy cast irons*.

catch plate (*mech*) A plate fitted to the spindle of a lathe in place of the chuck when turning between centres. The catch plate carries a peg or dog that engages with and drives a carrier mounted on the workpiece as shown in Fig. C.4.

cathode (*elec*) Negative electrode. **1** In electrolysis negative ions form at the cathode and positive ions are discharged. **2** In a thermionic device the cathode is the heated electrode from which

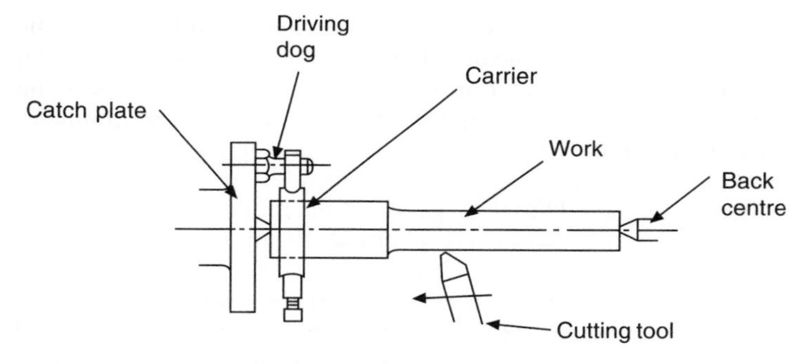

Fig. C.4 Catch Plate and Carrier

electrons are emitted into the interelectrode space. **3** In a semiconductor device it is the electrode *to* which the forward current flows. **4** In a thyristor it is the electrode *from* which the current flows.

cathode ray oscilloscope (CRO) (*elec*) See: *oscilloscope*.

cathode ray tube (*elec*) Device consisting of a heated cathode, called an electronic gun, contained within an evacuated glass tube. The electronic gun produces a beam of electrons by a process known as *thermionic emission*. This beam of electrons is accelerated down the tube and focused by a series of high-potential anodes. A deflection system consisting of two pairs of plates producing electric fields in vertical and horizontal planes controls the position of the beam before hitting a screen. The inside surface of the screen is coated with a phosphor which glows when electrons strike it, providing a visual display. This is the basis of radar viewing screens and the *cathode ray oscilloscope*. Television viewing screens are larger and electromagnetic deflection is used.

cathodic protection (*matls*) Protection of metal against electrolytic corrosion by making it the cathode in an electrolytic cell. Frequently used for protecting buried metal pipes that would normally be anodic relative to the surrounding moist earth, cathodic protection is achieved by means of an impressed e.m.f. which renders the pipe line to be negative relative to its surrounding environment, as shown in Fig. C.5.

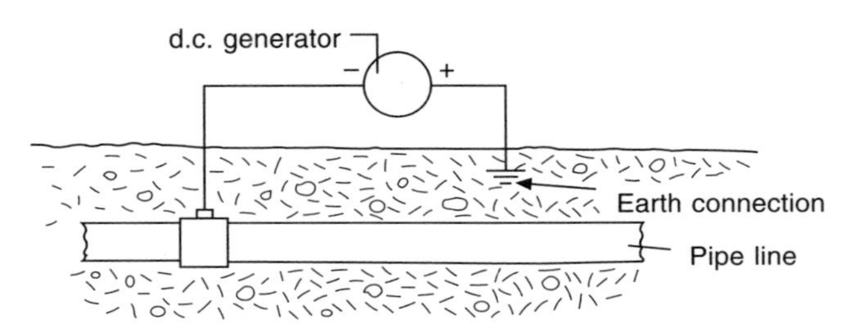

Fig. C.5 Cathodic Protection-impressed Current

caustic embrittlement (*matls*) The intercrystalline corrosion of steel in contact with hot caustic (alkaline) solutions, e.g. the corrosion of boiler shells.
CD-ROM (*comp*) An optically read disk used by computers for storing large quantities of digital data. As its name implies it is similar to the compact discs used for recording music digitally and it is used as a 'read only memory'. It is encrypted and read using laser light.
cell (battery) (*elec*) A device for generating an e.m.f. by two electrodes in contact with an electrolyte. **1** The chemical reaction may be spontaneous as in a voltaic cell (primary cell). This reaction cannot be reversed and the cell is discarded when the reactive chemicals are exhausted. **2** The chemical reaction may be the result of a current from an external source causing a chemical change within the electrolyte. The reaction can be reversed at a later time, releasing the electrical energy again as in an accumulator (secondary cell).

celsius heat unit (chu) (*sci*) The quantity of heat energy required to raise a 1 lb mass of pure water through 1 degree celsius.
celsius scale (*sci*) Temperature scale in which the reference points are the freezing point of water at 0°C at standard pressure and the boiling point of water at 100°C at standard pressure. This scale was formerly know as the *centigrade* scale. See also: *kelvin*.
cementation processes (*matls*) This is the generic name for three anti-corrosion treatments for ferrous metal products. (1) *Sherardizing* is the most widely used cementation process in which cleaned and pickled carbon steel products are placed in a rotating steel barrel along with zinc metal powder and heated to 370°C for up to 12 hours. The zinc film formed on the surface of the products provides an excellent 'key' for subsequent painting. (2) *Calorizing* is similar to sherardizing except that aluminium powder is used at a somewhat higher process temperature. Calorizing provides corrosion prevention against high-temperature corrosion (e.g. car exhaust systems), rather than against ambient-temperature corrosion for which sherardizing is more suitable. (3) *Chromizing* is similar to previous processes except that a mixture of aluminium and chromium powders is used and the processing temperature is at least 1200°C in a hydrogen atmosphere to prevent oxidation. This is an expensive process and is normally used for products to be used in hostile environments such as chemical plants. Chromizing must not be confused with *chromating*.
cementite (*matls*) The metallurgical name for the hard and brittle compound iron carbide found in steels and cast irons. The higher the cementite content, the harder and more brittle the metal becomes.
C. Eng. Abb. Chartered Engineer, a title which can only be used in the UK by engineers with acceptable qualifications and experience, registered with the Engineering Council.
centi (*sci*) Unit prefix meaning one hundredth, symbol c. See table of unit prefixes (Appendix 6).
Centigrade scale (*sci*) See: *Celsius scale*.

central processing unit (CPU) (*comp*) The 'brain' that controls the whole operation of a computer and can perform several functions. It fetches instructions in the form of programs from the memory, decodes the instructions and executes them. The CPU has an arithmetic logic unit that can perform instructed mathematical and logical operations. The CPU has an address counter that identifies the memory address of the next instruction to be processed in the correct order. It also has many registers for the temporary storage of data and bus control circuitry where the term 'bus' refers to the physical paths followed by data being fed into the CPU for processing or from the CPU to other elements of the computer after processing.

central tendency (*stats*) A single value, which is representative of a set of numbers, may be used to give an indication of the general size of the members of the set. The word 'average' is often used to indicate that single value. In statistics 'average' is the arithmetic mean (or just 'mean'). Other measures of central tendency are median and modal values. See: *mean*; *median*; *modal value*.

centre (*mech*) A cone-shaped device made of hard carbon tool steel for supporting work when turning in a lathe or other machine tool. The included angle of the cone is ground to 60°. When turning between centres on a lathe the ends of the workpiece need a 60° piloted centre hole cut into them to receive the centres; this is done using a *piloted centre drill*. The tailstock centre, or dead centre, is stationary whilst a live centre in the lathe spindle is designed to revolve with the workpiece. Sometimes a revolving centre mounted in ball or roller bearings is used in the tailstock when heavy loads and/or high-speed turning is required.

centre drill (*mech*) Tool used by a lathe or a drilling machine for drilling holes in the ends of a workpiece prior to turning between centres. The centre drill produces a 60° countersink with a small pilot hole in the bottom which ensures that the lathe centre supports the work on its conical surface and not on its tip.

centre lathe (*mech*) A versatile machine tool for producing surfaces of revolution and also plain surfaces. The workpiece is rotated and may be supported between two centres, gripped in a chuck or bolted to a face plate, and can be rotated at various speeds. When the lathe is for cutting metal, the cutting tool is held in a toolpost and controlled mechanically or automatically. Different profile tools may be used and the toolpost can be moved linearly at various speeds either parallel to the work to produce cylindrical components or at right angles to the saddle by traversing the cross-slide to produce plain surfaces. The cross-slide carries the compound slide which may be inclined to the axis of the workpiece for the purpose of producing tapered components. Operations such as screw cutting and drilling may also be carried out. Fig. C.6 shows the various features of a typical centre lathe.

centre of gravity (*sci*) The point within a body through which the total gravitational force acting on a body may be considered to act. A body behaves as if all of its mass were concentrated at this point.

centre-square (*mech*) An instrument for finding the centre line of the cross-section of a round blank. The centre-square is positioned over the blank with the V-location against the circumference of the blank. A line scribed along the straight edge of the centre-square will pass through the centre of the blank. The centre of the blank can then be found by scribing two or more evenly divided centre lines around the bar, the intersection of the lines being the centre of the blank.

centrifugal casting (*mech*) Casting process for cylindrical objects such as large-diameter water pipes and engine cylinder liners, where the mould is rotated so that the molten metal is spun out to exert a centrifugal force on the mould. The volume of metal poured into the mould controls the thickness of the casting, which will have a uniform wall thickness. No core is required.

centrifugal compressor (*mech*) A compressor functioning by

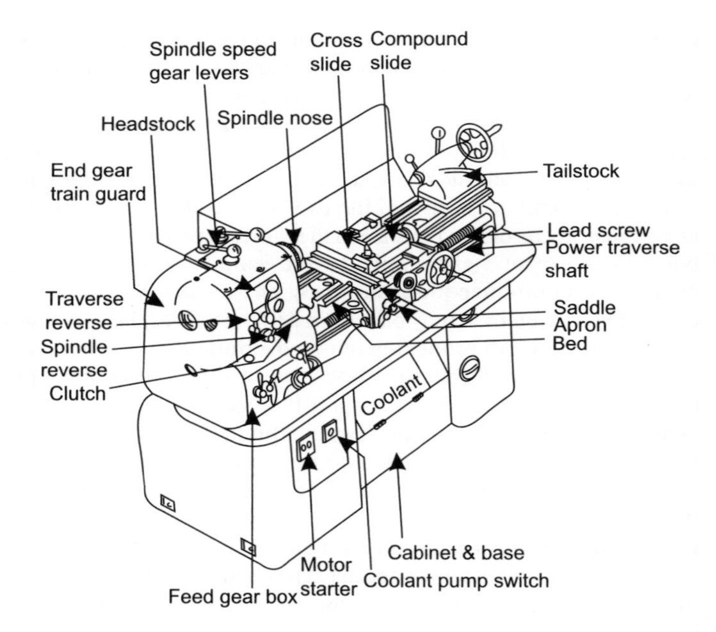

Fig. C.6 The Centre Lathe

passing air through a series of impellers, reducing the specific volume and increasing the pressure. See also: *centrifugal pump*.

centrifugal force (*sci*) The reaction force equal and opposite to the centripetal force acting on a rotating body. See: *Newton's third law*.

centrifugal pump (*mech*) A continuously acting pump with a rotating impeller that increases the pressure of a liquid by increasing the liquid velocity and applying a centripetal force. The velocity of the liquid is increased by the rotary action of the impeller and the kinetic energy is converted to pressure energy. The most common centrifugal pump is the *volute type*, so called because of the shape of the casing. The liquid is fed into the centre of the revolving impeller. Centrifugal action causes the liquid to flow radially through the impeller to the pump casing which is snail-shaped and designed to convert the kinetic energy into pressure energy. The volute type of centrifugal pump is used where a high flow rate is required against a low head. The second type of centrifugal pump is the *diffuser type*. This consists of a rotating impeller as previously, but the kinetic energy is converted to pressure energy by a diffuser ring of fixed blades which form diverging passages. The diffuser type of pump is suitable for pumping against a high head sometimes arranged in several stages, e.g. for a boiler water feed.

centripetal force (*sci*) The force acting on a body causing it to move in a circular path. If an object tied to a piece of string is swung in a circular path, the inward pull of the string on the object is referred to as the *centripetal force*. Although the body may be travelling in a circular path with a constant velocity, as the direction is constantly changing there is an acceleration towards the centre of the circular path. This acceleration is equal to $\omega^2 r$ where ω is the angular velocity in rad/s and r is the radius of the circular path in metres. The force that the object exerts on the string in attempting to travel in a straight line is referred to as the centrifugal force. This has the same magnitude as the centripetal force but acts in the opposite direction. Centrifugal and centripetal forces are a major consideration in the design of rotating components of machinery and in the study of vehicle dynamics.

centroid (*maths*) Centre of volume or centre of area of a shape. If it is used to estimate the centre of gravity of a mass, a uniform density of material must be assumed; the centre of area must assume a uniform density and a uniform thickness of material.

ceramics (*matls*) General classification of inorganic, non-metallic materials that are processed at, and may be used at, high temperatures. They consist mainly of silicon chemically combined with non-metallic elements such as oxygen, carbon and nitrogen. Metal compounds such as metal oxides may also be present. Ceramics are used in engineering for a wide range of products including cutting-tool materials (metallic carbides and oxides), piezoelectric transducers, insulators, magnets, refractories and fibres for reinforcement and optical data transmission. Their usefulness in engineering applications lies in their ability to maintain their strength at high temperature, hardness and resistance to wear.
chain block (*mech*) Pulley system mechanism using chains instead of ropes for lifting heavy loads by hand.
chaplet (*mech*) Iron support for the core of a sand mould to prevent the core from being distorted or displaced by the inflow of molten metal when casting.
character (*CAE*) A character is a number, letter or symbol that is recognized by a CNC machine's control unit.
character code (*comp*) Binary code that represents characters from a computer keyboard, e.g. an *ASCII* code.
charge (*elec*) **1** Electric charge is a quantity of electricity equal to a flow of one ampere of current for one second; quantity symbol Q and unit the coulomb (C). **2** When associated with storage batteries (accumulators, secondary cells) a larger unit of charge called the ampere-hour (Ah) is used. **3** (*mech*) The amount of material or work loaded into a furnace.
Charles' law (*sci*) A law that applies to perfect gases and states that for a mass of gas held at a constant pressure the volume is directly proportional to the temperature ($V \propto T$). This relationship was established by J.A.C. Charles and J. Gay-Lussac in the late eighteenth century. See also: *gas laws*.
Charpy impact test (*mech*) An impact test to determine the toughness of materials. A notched test piece is supported at each

end as a beam as shown in Fig. C.7, rather than held as a cantilever as in the *Izod test*. The impact load can either be 150J or 300J and the test piece is struck at its centre, opposite the notch and equidistant from each support. The energy absorbed in bending or breaking the specimen is a measure of its toughness. See: *Izod test*.

chaser (*mech*) **1** Centre lathe: internal or external lathe tools with an edge that profiles a specific screw thread. They may be flat or circular and can be mounted on a centre lathe toolpost.

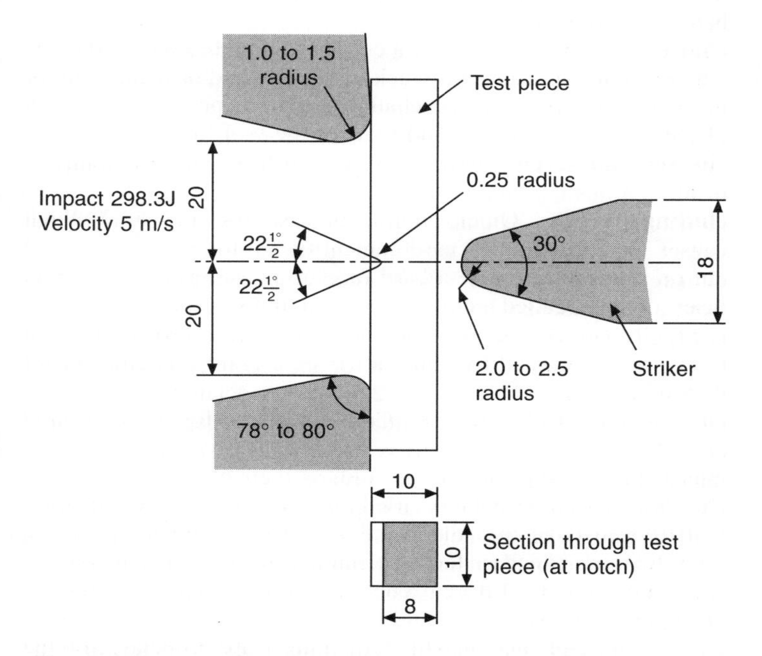

Fig. C.7 Charpy Test (all dimensions in millimetres)

Chasers are used for giving an accurate profile to screw threads that have been roughed out using a single-point tool. Initial roughing out is not necessary for fine-pitch threads. **2** Capstan, turret and automatic lathes: these machines cut threads on the ends of components using die-heads containing four chasers equally spaced around the circumference of the thread. The chasers may be arranged radially or tangentially. A single chaser may be mounted on the cross-slide toolpost of these machines when cutting threads whose diameter precludes the use of die-heads or where the thread is behind a shoulder.

chatter (*mech*) Vibration of a cutting tool and/or the workpiece when machining, caused by lack of rigidity and/or inadequate or incorrect support. Excessive chatter results in a poor-quality finish of the machined surfaces and reduces the tool life.

cheese-head screw (*mech*) Screw with a slotted cylindrical head.

chilling (*mech*) Quick cooling of cast iron and steel which causes the formation of hard cementite. Chilling can be carried out on purpose to create a hard face of a casting, by putting an insert of steel, called a chill, into the mould so that molten metal is rapidly cooled and becomes hard. This can also be done to relieve stress caused by some areas of a casting cooling faster than others.

chip or silicon chip (*elec*) Common name for an *integrated circuit*.

chisel (engineers) (*mech*) **1** Cold: as their name implies, cold chisels are roughing-out tools used for cutting metal by hammering a sharpened edge into the metal at normal room temperature. They are made of carbon tool steel, hardened and tempered. Various types are used for different cuts and different metal forms, for example a wide surface would be broken down into strips using a cross-cut chisel, and then the remaining strips of metal removed using a flat chisel. **2** Hot: these are used by blacksmiths for cutting

metal heated to or above red heat. They are similar to cold chisels but are provided with a handle perpendicular to the chisel and have a less-acute cutting edge. See: *Hardie*.
choke (*elec*) Popular name for an electrical *inductor*.
chopper (*elec*) **1** Electronic circuit for converting d.c. to a.c. by rapidly switching a voltage source on and off to produce a rectangular waveform varying between zero and the full potential of the d.c. input. **2** A light interrupter to control the electrical output of a photoelectric cell.
chromating (*matls*) This process is used to protect magnesium alloy castings which are dipped into a hot solution of potassium dichromate to form a hard and protective oxide film, which is then sealed with a zinc chromate paint after which it is finished with a decorative paint system. Chromating must not be confused with *chromizing*.
chromizing (*matls*) See: *cementation processes*.
chuck (*mech*) Device for gripping drills usually below 15 mm diameter and also the workpiece of a lathe. **1** Drill chucks: these have three adjustable, radial jaws that clamp on the shank of the drill. **2** Three-jaw, self-centring chucks: these are used on centre lathes for holding work with cylindrical or hexagonal surfaces. The jaws are adjustable radially by means of gears and a scroll that engages with teeth on the back of the jaws and causes the jaws to move concentrically with the spindle axis of the machine. There are separate sets of internal and external jaws and these must be kept and used only for the chuck with which they have been supplied. They are also numbered so that they can be inserted into the chuck body in the correct sequence. **3** Four-jaw independent chucks: These have four reversible jaws that can be used externally or internally. They move radially in the chuck body and each jaw has its own clamp screw, which allows the jaws to be moved independently of each other. Four-jaw chucks can be used to hold cylindrical work, rectangular work and work that has to be offset

for eccentric turning. They are more difficult to set up than self-centring chucks but are more versatile as the jaws operate independently and grip more efficiently.

circlip (*mech*) Spring clip in the shape of a 'C', used for fixing components in place; internal circlips fit in a groove in a hole, and external types fit on a grooved shaft.

circle (*maths*) A closed curved shape that maintains a fixed distance, known as the *radius*, from a fixed point within the curve, known as the *centre*. The perimeter is known as the *circumference*. The *diameter* is a line that passes from one point on the circumference, through the centre of the circle, to the circumference at the other side, and is always twice the length of the radius. The length of the circumference divided by the length of the diameter is a constant of approximate value 3.141 59, which is referred to as π (the Greek letter pi); algebraically, $C = \pi d$ where C is the circumference and d is the diameter of the circle, or $C = 2\pi r$ where r is the radius. The area of a circle, $A = \pi r^2$. As the diameter is generally easier to measure in engineering than the radius, the more common form of the formula is $A = \frac{\pi d^2}{4}$. A semicircle is half a circle and a quadrant is a quarter of a circle. The parts of a circle are shown in Fig. C.8.

circuit 1 (*elec*) A closed path of conductors through which an electric current or signal may flow. **2** (*mech*) More generally, a closed path for any medium such as hydraulic oil or heating water.

circuit-breaker (*elec*) Device designed for opening an electric circuit either manually or automatically for safety purposes, e.g. owing to excessively high current. See also: *overcurrent relay*; *fuse*.

circulating pump (*mech*) **1** A centrifugal pump used in a liquid circuit for circulation, e.g. the water pump of a hot water heating system. The use of a centrifugal-type pump ensures that the flow

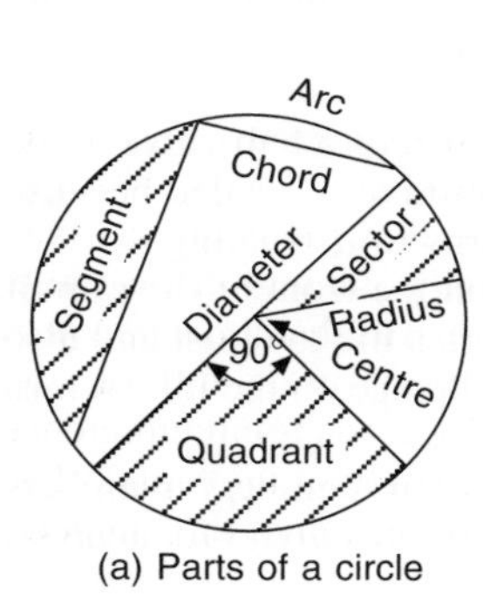

(a) Parts of a circle

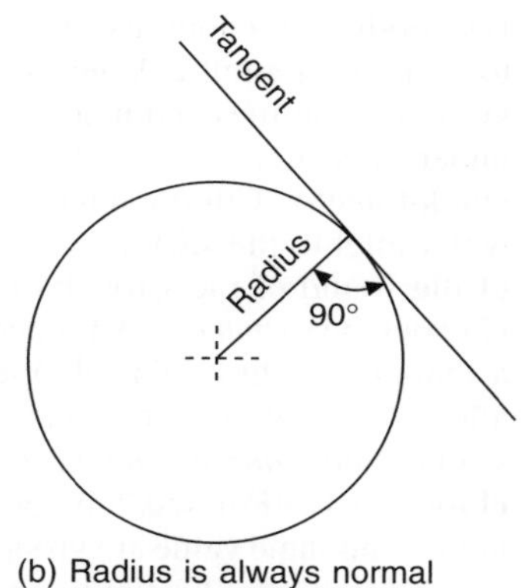

(b) Radius is always normal (90°) to its tangent

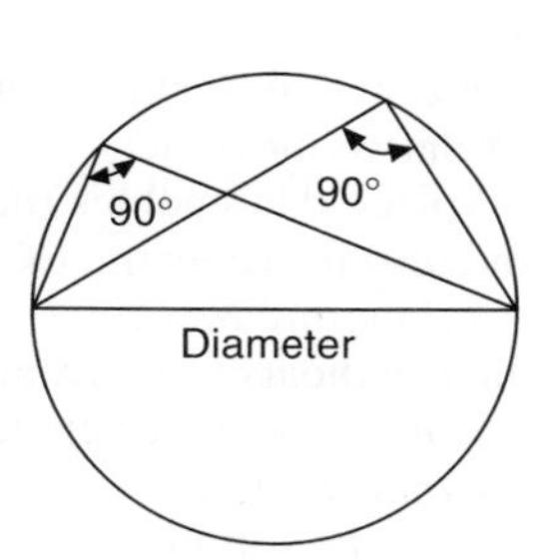

(b) Angle in a semi-circle is always 90°

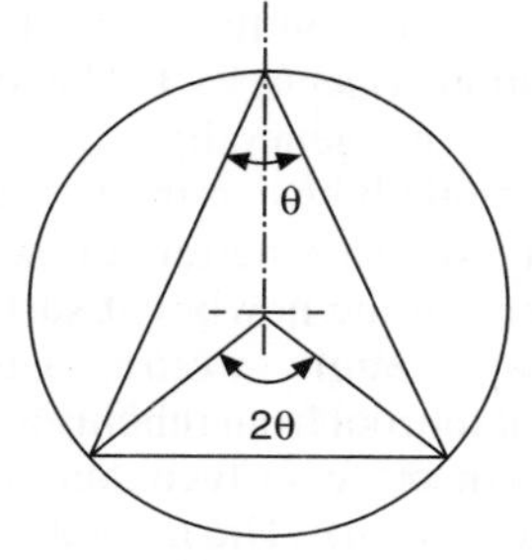

(b) Angle subtended at the centre of a circle is always twice the angle at the circumference

Fig. C.8 Parts and Properties of a Circle

is smooth, continuous and free from pulsations. Sometimes referred to as a circ. pump. **2** A gear-type pump is used in the lubricating system of engines and machine tool gearboxes to circulate the oil under pressure.

clack (*mech*) Common name for the non-return ball valve at the water inlet to the steam boiler of a locomotive, so called because of the sound of the valve ball as it returns to its seating.

clapper box (*mech*) A hinged device on the front of the ram of a shaping machine. The clapper box carries the toolpost and also allows the tool to swing clear of the workpiece on the idle, return stroke. See: *shaping machine.*

class (*stats*) For sets having more than ten members, members having the same value are grouped together into *classes* for analysis into a *frequency distribution* table.

claw-clutch (*mech*) A shaft coupling comprising an axially toothed flange which slides along a shaft to engage with another similar flange on the other shaft. Also called a dog-clutch.

clearance (*mech*) **1** The distance between two surfaces or components, generally required to allow relative movement. **2** A relief angle between the tool and the workpiece that enables the cutting tool to penetrate the workpiece and also prevents the tool rubbing on the newly cut surface. See: *metal cutting tool angles.*

clearance angle (*mech*) A tool angle that prevents the cutting edge of the tool from rubbing and allows the cutting tool to penetrate the work piece surface. See: *metal cutting tool angles.*

clinker (*mech*) The incombustible remains of the solid fuel used in coal-fired boiler is normally in the form of a powdery or granular *ash.* However, when using some grades of coal, under high-temperature conditions, the ash will sometimes fuse into a solid mass called *clinker.* This may be used for the manufacture of load-bearing thermal insulation blocks for building purposes.

clipboard (*comp*) Temporary storage area of a computer operating system, used to transfer information from one application program or document to another.

clock (*comp*) Electronic timing device of a computer system for synchronizing operations.
closed cycle (*sci*) A circuit of open systems through which the working fluid circulates continuously through a cycle of mechanical and thermodynamic states. See also: *thermodynamics—first law*.
closed loop (*elec/mech*) A feedback control system that includes a process output measurement system, and allows a feedback signal to be compared by a controller with a desired value; so called because a signal can travel around a loop to return to its starting point. The aim of the controller is to maintain the measured variable at the process output as closely as possible to the desired value. The measured variable provides the feedback signal that is compared with a desired value to produce an error signal. The control system should reduce the error to zero whenever there is a change in the desired value or a system disturbance. Some systems can include many loops. See also: *feedforward*, *automatic control system*.
closed system (*mech*) Term given to a thermodynamic system in which no matter crosses the boundary during the process under investigation. Heat and work may cross the boundary, but the same matter remains within the system.
clutch (*mech*) A device for connecting one driveshaft to another shaft or wheel. **1** Friction clutch: this is used to provide progressive engagement between the drive motor and the transmission system of machine tools when a smooth and controlled take up of the drive is required. **2** A dog- or claw-clutch provides permanent engagement or disengagement and cannot provide a progressive take-up of the drive. No friction is involved, and the teeth on one element of the clutch are either in total engagement or total disengagement with the teeth on the mating clutch element. The main advantage of this type of clutch is its simplicity and that in no way can it slip; the drive is positive.
CMOS (*elec*) Complementary metal oxide semiconductor logic,

contemporary integrated circuit technology with typical characteristics of: fan-in = 8, fan-out = 50, propagation delay = 30 ns, noise margin = 1.5 V, power consumption = 1 mW. The low power consumption of this technology makes it particularly useful for battery-powered equipment.

coach bolt (*mech*) Bolt with a rounded head and a square section under the head that fits into a corresponding depression in the joining material, to prevent the bolt from turning when the nut is tightened. It is also called a cup-square bolt.

COBOL Abb. (*comp*) Common business oriented language; high-level computer language used in business data processing.

coefficient of contraction (*sci*) When applied to the study of a small sharp-edged orifice in the side of a tank containing liquid, the coefficient C_c is the ratio of the cross-sectional area (C.S.A.) of the fluid jet through the orifice at the *vena contracta* to the cross-sectional area of the orifice:

$$C_c = \frac{\text{C.S.A. at vena contracta}}{\text{C.S.A. of orifice}}$$

A typical value is 0.64. See: *vena contracta*.

coefficient of discharge (*sci*) Coefficient, C_d, applied to the study of a small sharp-edged orifice in the side of a tank of liquid, i.e. the ratio of the actual volumetric discharge to the theoretical volumetric discharge.

$$C_d = \frac{\text{actual discharge}}{\text{theoretical discharge}}$$

This can also be related to the *coefficient of contraction*, C_c, and *coefficient of velocity*, C_v:

$$C_d = C_c \times C_v$$

A typical value is 0.61.

coefficient of friction (*sci*) The ratio (μ) between the frictional force and the normal reaction when two dry, flat surfaces move or tend to move across one another. Expressed mathematically:

$$\mu = \frac{F_f}{R_N}$$

where F_f is the frictional force and R_N is the normal reaction of one surface on another; for an object resting on the surface of the Earth, R_N will be equal to the weight of the object. This ratio can be calculated either as the coefficient of *static friction* (limiting friction) at the limiting point of friction where movement is about to commence, or as the coefficient of *kinetic friction* when the two surfaces are sliding across one another. See also: *friction*.

coefficient of linear expansion (*sci*) The thermal coefficient of linear expansion (α) of a substance can be defined as the change in length of a unit length of that substance when its temperature is changed by one degree. Expressed mathematically, the linear expansion of a solid during a temperature change is as follows:

$$x = l \cdot \alpha \cdot \delta\theta$$

where x is the magnitude of linear expansion or contraction, l is the original linear dimension of the material, α is the coefficient of linear expansion and $\delta\theta$ is the temperature change.

coefficient of performance (*sci*) The performance criterion relating heat transfer to work input for a cycle. For a refrigeration cycle the coefficient of performance is the ratio of the heat transferred at the low temperature to the work input. For a heat pump, the coefficient of performance is the ratio of the heat transferred at the high temperature to the work input.

coefficient of restitution (*sci*) A coefficient that takes account of the level of elasticity of a body. It is represented by e and has a value between 0 and 1. For a perfectly elastic collision between two bodies $e = 1$; for a perfectly non-elastic collision $e = 0$ and there is no rebound. It can be used in collision calculations as follows:

relative velocity after impact

= relative velocity before impact × (–*e*)

coefficient of thermal expansion (*sci*) A constant used for predicting how much a substance will expand or contract with a change in temperature. For solids, the coefficient of linear expansion predicts the change in length of a linear dimension; for areas the coefficient of superficial expansion is used and for volumes the coefficient of cubical (volumetric) expansion is used; for liquids the coefficient of volumetric expansion is used. See: *coefficient of linear expansion*; *coefficient of volumetric expansion.*

coefficient of velocity (*sci*) The coefficient of velocity, C_v, as applied to the study of a small sharp-edged orifice in the side of a tank or reservoir of liquid, is the ratio of the actual fluid velocity through the orifice at the *vena contracta* to the theoretical velocity

$$C_v = \frac{\text{actual velocity}}{\text{theoretical velocity}}$$

A typical value is 0.96.

coefficient of volumetric expansion (*sci*) The volumetric expansion of a liquid or solid during a temperature change can be calculated as follows: $\delta V = \gamma \cdot V \cdot \delta T$ where δV is the magnitude of volumetric expansion or contraction, V is the original volume of the liquid or solid, γ is the *coefficient of volumetric expansion* and δT is the temperature change.

coercive force (*elec*) The amount of reverse magnetic field strength required to reduce the residual flux of a magnetic material to zero. The coercivity of a material is a measure of the difficulty in restoring the residual flux to zero. See also: *hysteresis.*

coercivity (*elec*) See: *coercive force.*

coil (*elec*) Insulated wire wound in a coil, possibly around a former; such a coil is also called a *solenoid*; current passing through a conductor produces an electromagnetic field around the conductor and, by winding the conductor into a coil, that electromagnetic field is concentrated. It may be concentrated still further by inserting a *core* of *ferromagnetic* material into the centre of the coil to form an electromagnet.

coining (*mech*) Pressing operation where a relatively thick metal blank is trapped between dies carrying an engraved, negative impression. The metal is forced to flow into the die impressions so as to leave a raised (positive) impression on both sides of the blank, as in a monetary coin from which the process gets its name. The process is usually carried out 'cold' (at room temperature) but can be performed 'hot' to relieve the pressure on the dies. See: *embossing*.

combined gas law (*sci*) This is a combination of Charles's law and Boyle's law into a combined equation as follows: $(p_1V_1)/T_1 = (p_2V_2)/T_2 = (p_3V_3)/T_3$ etc. See: *gas laws*.

combined heat and power units (CHP) (*sci*) A generating plant with a waste heat recovery system that can be applied to a useful heat load, usually a central-heating system. The waste heat is recovered by heat-exchangers. Heat from the exhaust gases is transferred to the engine coolant as it leaves the engine via the primary heat-exchanger. Heat from the coolant is then transferred to the heat load by the secondary heat-exchanger. A well-maintained CHP unit can operate at 80–90% efficiency and can reduce bills compared with the conventional method of purchasing electricity from the national grid and supplying heat from a boiler plant. There are also environmental benefits in using CHP units rather than conventional energy generation methods.

common logarithm (*maths*) See: *logarithm*.

common-mode rejection ratio (*elec*) Ratio of the gain for a differential signal to the gain for a common-mode signal of the same amplitude.

common-mode signal (*elec*) Signal applied to both inputs of a differential amplifier.

commutation (*elec*) The method by which connections are made to the rotating conductors of an electrical generator or motor, using brush contacts through which the current flows. See: *commutator*.

commutator (*elec*) A device for conveying electric currents to or from the rotating conductors of the armature of an electrical machine such as a motor or generator. Generally associated with d.c. machines, it is also used in universal fractional horsepower motors designed to run on d.c. or a.c. Such motors are frequently found in vacuum cleaners and hairdryers. In the case of d.c. generators (dynamos) it coverts the alternating currents induced in the armature coils into direct current at the output of the machine. The commutator is mounted on the same shaft as the armature and rotates with it.

compensation (*elec*) The fitting of an electrical winding or other device in an electrical circuit to compensate for the effects of current flow or to adjust the distribution of magnetic flux; for example, a compensating winding may be added to a series motor to counteract the effect of armature reaction and improve the power factor.

compiler (*comp*) An interpreter program for a microcomputer that translates statements made in *a high level language*, such as Pascal, into *machine code* for loading into the memory and executing.

complex number (*maths*) A number y of the form $a + \mathbf{i}b$ where $\mathbf{i} = \sqrt{-1}$, and a and b are real numbers. The number a is the real part and is written Ry and b is the imaginary part and is written Iy. Thus if $y = 2 + 3\mathbf{i}$, then R$y = 2$ and I$y = 3$. There can be no real roots for $\sqrt{-1}$ and **i** is called an operator in pure mathematics. In electrical engineering **j** is used instead of **i** to avoid confusion with the symbol for current.

composite material (*matls*) Material made by reinforcing one material using the fibres or particles of another dissimilar material. For example, glass-fibre–reinforced polyester mouldings.
compound bar (*matls*) A structural component made of two or more materials, arranged in parallel, rigidly connected together at the ends with an axial load. Generally, this is an ideal arrangement to simplify the study of stress and strain of more complex components made of two or more materials.
compound gear train (*mech*) An arrangement of several gear trains used in series, with the driven gear wheel of one train fixed onto the same shaft as the driver of the next, as shown in Fig. C.9.
compound slide (*mech*) The top slide of a centre lathe mounted on the cross-slide and carrying the toolpost. The compound slide can be set to any angle relative to the workpiece axis and is able to move the tool a short distance in that direction. It is used for cutting short tapers such as chamfers. For longer tapers a taper turning attachment may be used or the tailstock can be offset when turning between centres. See: *taper turning attachment*.
compound wound generator (*elec*) A generator with both series and shunt field windings, that is, field windings in series with the armature windings and field windings in parallel with the load. As the load on the generator becomes greater the current through the series-connected field coil becomes greater, increasing the magnetic field and helping to prevent the drop in potential at the output of the machine that would otherwise occur owing to the internal resistance of the armature windings.
compressed air (*mech*) Air at higher than atmospheric pressure. It is used for the transmission of energy in situations where electrical motors would be dangerous, or for control actuation purposes. See: *air compressor*.
compression ratio (*sci*) Ratio of the maximum volume of the cylinder of an internal combustion engine when the piston is at

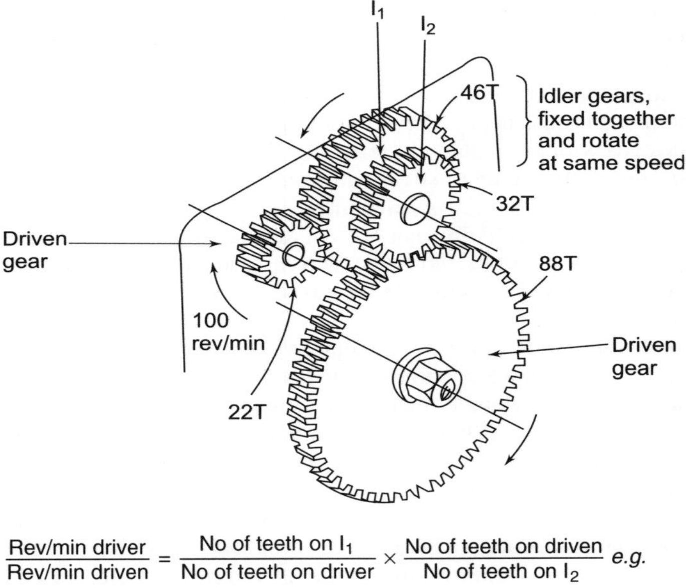

$$\frac{\text{Rev/min driver}}{\text{Rev/min driven}} = \frac{\text{No of teeth on } I_1}{\text{No of teeth on driver}} \times \frac{\text{No of teeth on driven}}{\text{No of teeth on } I_2} \quad \textit{e.g.}$$

$$\frac{100 \text{ rev/min}}{\text{Rev/min driven}} = \frac{46t}{22t} \times \frac{88t}{32t}$$

$$\text{Rev/min driver} = 100 \times \frac{22}{46} \times \frac{32}{88} = 12^1/_2 \text{ rev/min}$$

Fig. C.9 Compund Gear Train

bottom dead centre to the minimum cylinder volume when the piston is at *top dead centre*. The maximum volume is the sum of the swept and the clearance volume.

$$\text{compression ratio} = \frac{\text{swept volume + clearance volume}}{\text{clearance volume}}$$

Petrol engines have compression ratios between 71:1 and 101:1 whereas diesel engines require much higher compression ratios of between 151:1 and 241:1.

computer (*comp*) Device that can process data in a way specified by instructions written in a program of arithmetic equations. The most commonly used type is a digital computer where characters are represented by binary words and comprise an input device such as a keyboard, a CPU for controlling and performing operations, memory for storage, for example magnetic disks, and an output device such as a *VDU* and a printer. See also: *hardware*; *software*.

computer-aided design (CAD) (*mech*) The use of computers with high-resolution graphics together with specialized software packages for the drawing and design of products ranging from integrated circuit chips to supertankers. Designs can be modified, updated and evaluated quickly and easily compared with manually prepared drawings. Layering techniques (colour-coded) can be used to simplify buildings and plant layout drawings; for example, one layer can show the layout of the building, another layer shows the electrical services, another layer the plumbing services, another layer air-conditioning services, etc. For small components it is also possible to link the digitized design directly to a CNC machine for prototype production by using suitable post-processor software. See: *CAD/CAM*.

computer-aided manufacture (CAM) (*CAE*) Computer software systems such as PEPS (production engineering productivity system) that can import *DXF* files. These, in turn, can be defined as part boundaries (outer profiles and inner pockets and slots called K curves). The language used in such systems is very powerful with one-line commands defining pockets and profiles. Once the program has been proved, by simulation, that the cutter path is correct, then the simulation can be post-processed and downloaded to the controller of suitable CNC machines for immediate production. See: *CAD/CAM*.

Computer numerical control (CNC) (*CAE*) The control of machine tool functions by actuators driven by a dedicated computer. All the necessary data to control the machine and produce the required component is encoded in a language that the computer built into the machine controller can read. At one time the program was produced manually, a skilled and tedious process, and then encoded on a punched tape for loading into the machine. Nowadays computer-aided programming is used that automatically does all the calculations and can also prove the program by simulation before downloading it into the machine. The simulation ensures that the cutter path is correct and that the cutter will not collide with any clamping devices. In-cycle gauging and other facilities ensure that parts produced on CNC machines have a consistent quality and a high order of accuracy.
concentricity (*mech*) When two components have a common axis they are said to be *concentric*. The accuracy of alignment of their individual axes being a measure of their concentricity. For example, the bore of a wheel should be concentric with its circumference. See also: *eccentricity*.
concurrent system of forces (*sci*) System of forces that pass through the same point. A concurrent system of forces has no moment and does not introduce any turning effects.
condensate (*sci*) The liquid formed as a result of the *condensation* of a vapour by cooling.
condensation (*sci*) The formation of a liquid from its vapour.
condenser (*sci*) **1** Heat-exchanger unit kept at a partial vacuum for condensing the exhaust steam of a steam power cycle using cooling water. **2** Device of refrigeration cycle for transferring the heat from the refrigerant to the surroundings. **3** (*elec*) Old name for a *capacitor*.
conductance (*elec*) See: *conductivity*.
conduction 1 (*elec*) Electrical conduction is the flow of electric charge through a material owing to an electromotive force. See

also: *electric current*. **2** (*sci*) Thermal conduction is the transmission of heat through a substance from a high temperature to a lower temperature. All materials can conduct heat in any state but this is usually only significant in solids, where the transmission of energy is partly due to the impact of adjacent molecules and partly to internal radiation. When the solid is a metal, there are also a large number of freely moving electrons which transmit the majority of the energy through migration; this explains why good thermal conductors are also good *electrical conductors*.

conductivity 1 (*elec*) Electrical conductivity, or conductance (G), is a measure of the ease with which an electric current can flow through a material when an electromotive force is applied. It is the reciprocal of *resistance* and is measured in siemens (unit symbol S):

$$G = \frac{1}{R} \text{ siemens}$$

2 (*sci*) Thermal conductivity is a measure of a material's ability to conduct heat. One-dimensional steady heat flow, e.g. through a wall with a temperature gradient across the faces, can be estimated using Fourier's law:

$$\dot{Q} = -k \cdot A \cdot \frac{\Delta T}{x}$$

where $\dot{Q}$ is the heat flow rate through the material measured in W or J/s, k is a constant known as the thermal conductivity of the material measured in W/mK, A is the cross-sectional area of the flow measured in m^2, ΔT is the temperature difference across the material measured in K and x is the thickness of the material measured in m. The negative sign indicates that heat flow is positive in the direction of a temperature decrease. The relationship assumes that the two faces are parallel and that all the heat transfer is normal to the faces, none being lost through the sides of the material.

conductor (*elec*) **1** A material that has a high electrical *conductivity* and is used to conduct electric currents. Most metals are good electrical conductors and most non-metals (with the exception of carbon) are very poor conductors (insulators). See also: *insulator*. **2** (*sci*) Material that has a high thermal *conductivity*. Most metals are good thermal conductors owing to the large number of freely moving electrons they contain.

conduit (*elec*) Protective trunking or tube for carrying electrical cables.

cone drive (*mech*) A belt drive system with several different-diameter pulley wheels on the same axis in a cone shape on one spindle driving an identical set of wheels facing in the opposite direction on a parallel spindle. Used in low-power workshop machinery, such as bench drilling machines, to obtain a range of speeds from a single-speed motor. See: *drilling machine*.

conic sections (*maths*) These are the curves that are produced when a plane cuts a right-cone in various ways, as shown in Fig. C.10. These curves can be described algebraically. For example, a parabola is the curve of any second-order (quadratic) equation.

connecting rod (*mech*) Component of a mechanism that connects a sliding member such as a piston to a rotating member such as a crank, thus converting linear motion into rotary motion or vice versa. Depending on the direction of the forces involved, the connecting rod can be in compression and acting as a strut or in tension and acting as a tie. The end of the connecting rod coupled to the sliding member is usually called the *little-end* and the end of the connecting rod coupled to the crank is usually called the *big-end*.

connective heat transfer coefficient (*sci*) Coefficient used for calculating the heat transfer rate through a fluid adjacent to a heat transfer surface; it assumes that the heat flow is proportional to

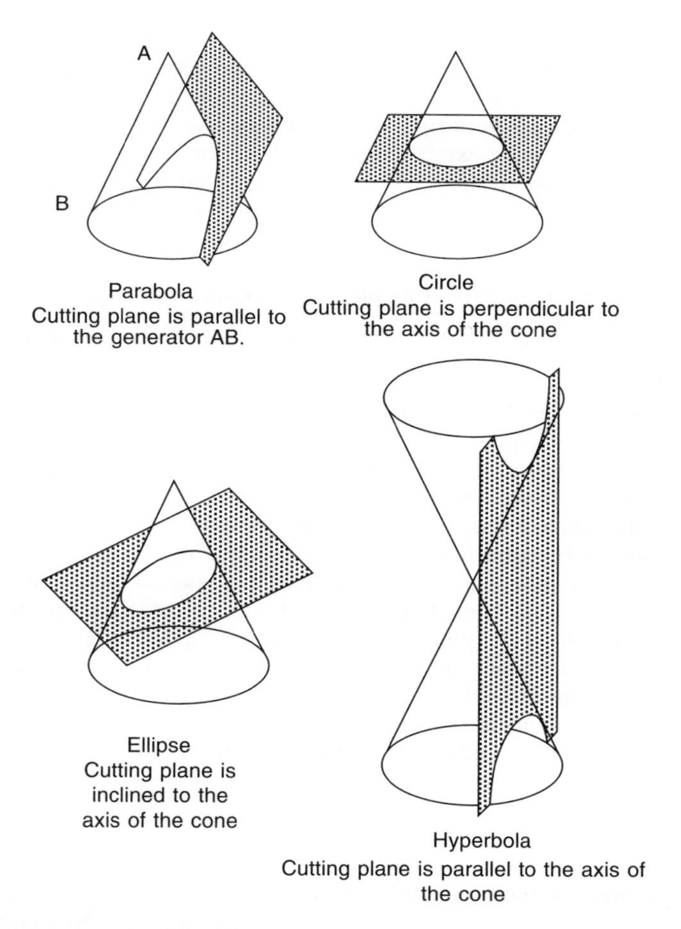

Fig. C.10 Conic Sections

the temperature gradient between the surface material and the fluid temperature at some distance from the wall; often represented by the symbol h.

conservation of energy (*sci*) The principle of the conservation of energy can be expressed by the statement 'energy can neither be created nor destroyed but can be converted from one form into another'. Energy cannot be lost but, in practice, it is sometimes considered lost when it is converted into a form that is not useful, e.g. the heat from the exhaust of an internal combustion engine. In thermofluids this principle can be written as:

energy in – energy out = change in energy of a system

this is the basis of the first law of thermodynamics.

conservation of momentum (*sci*) The principle of the conservation of momentum can be expressed by the statement 'momentum can only be destroyed by a force and only created by the action of a force'. If no external force acts on a body or a system then the momentum of that system remains constant.

continuity equation (*sci*) A law applied to liquids flowing through a pipe that is not of a constant cross-sectional area. This law states that 'the velocity of the flow is inversely proportional to the area of the pipe section'. Consider liquid flowing through a pipe that reduces in diameter over its length between points 1 and 2. The mass flow rate past any section is constant. If cross-sectional areas at sections 1 and 2 are A_1 and A_2, and the fluid velocities are c_1 and c_2 respectively, then

$$\frac{c_1}{c_2} = \frac{A_1}{A_2}$$

This is also known as the equation of continuity.

continuous casting (*mech*) A casting process, in which the molten metal is added at the top of a water-cooled mould and the solidified metal forms a plug at the bottom of the mould. As the metal solidifies the ingot so produced is extracted from the bottom of the mould as a continuous cast bar and very little scrap is produced.

continuous data (*stats*) Measured data that can have any value within prescribed limits is said to be *continuous*.

continuous path system (*CAE*) This is the most widely used system. The path taken by the tool in moving from one point to the next and the traverse rate is fully under the control of the programmer. Angular and curved paths can be taken in two or three dimensions simultaneously and complex profiles and contoured components can be generated. See: *parallel path system*; *point-to-point system*.

continuous signal (*elec*) Signal that exists at every instant in time, often referred to as an analogue signal. Compare: *discrete signal*.

contraction rule (*mech*) A measuring instrument used by pattern makers in which the distances between the conventional markings have been extended so that the pattern is made oversize. This allows for metal shrinkage as the casting solidifies and cools to room temperature. A separate rule is required according to the metal being cast.

control effort (*elec*) The output signal of the control unit of a control system. This signal is applied to the plant actuators in an attempt to maintain the output of the plant at a desired value set at the control unit (controller). See: *automatic control systems*; *closed loop*.

controlled input (*mech*) Plant inputs that are regulated by a control system in order that the plant can comply with the prescribed performance of requirements. See also: *disturbance inputs* and *automatic control systems*.

control system (*mech*) A pneumatic or electrical system that controls a plant or process. See: *automatic control systems*.

convection Thermal convection is a form of heat transfer that occurs owing to the movement of a fluid. In natural convection, fluids move owing to the changes in density that occur with changes in temperature. As a fluid increases in temperature, it expands and its density reduces. As a result, it will tend to rise by being displaced upwards by colder fluid that has a higher density. This process is employed in most central heating systems to assist the water flow around the system. In forced convection, the fluid movement is due to a pressure differential created by a pump. An example is the cooling system of a liquid-cooled internal combustion engine, where heat must be removed from the engine and transferred to the radiator. The pump creates a pressure increase and forces the coolant to circulate around the hot engine, where heat is transferred to the coolant. The coolant then flows around the system to the radiator tubes where heat is transferred to the circulating air, before returning to the pump.

conversational data input (*CAE*) The program is loaded into the control console by manually depressing keys. However, instead of writing out the program in machine code in advance of entering the data, the program is entered in response to questions (prompts) appearing on a visual display unit (VDU) in everyday, conversational English. The machine's computer is pre-programmed with standard data stored in 'files' in its memory. These can be called up into the program as required by the operator in response to a prompt. To reduce idle time, modern conversational control units allow a new program to be entered whilst an existing program is operating the machine. See: *manual data input*; *punched tape input*; *magnetic tape input*; *direct numerical control (DNC)*.

conversion coatings (*matls*) Phosphating processes used for corrosion prevention, previously known by such names as *bonderizing*, *granodizing*, *Parkerizing*, and *Walterizing*, have now been standardized under BSS 3189 and are collectively known as *conversion coatings*. The components to be treated are degreased and acid-pickled, after which they are immersed in a hot phosphating solution. The metal surface is converted into complex metal phosphates (most commonly used), chromates or oxides depending on the process used. Finally, the treated surface is sealed by oiling, waxing or painting. Such processes are widely used as a pretreatment for painting as they provide a 'key' for the paint and prevent corrosion under the paint film. Since phosphates have a high lubricity, phosphate conversion is frequently applied to metal surfaces as an extreme pressure lubricant.

converter (*matls*) See: *Bessemer converter*.

conveyor (*mech*) A mechanical handling device for moving materials, components or assemblies from one place to another, usually in discrete manufacturing processes.

cooling tower (*mech*) Structure used at power stations to cool water by an updraught of air whilst the water, admitted at the top of the tower, is broken down into droplets as it drips down over a framework which increases the surface area exposed to the air. The water so cooled is returned to the condensers where the exhaust steam from the turbines is converted back to water and reused in the boilers.

cope (*mech*) The top half of a moulding box for casting when sand moulding.

coplanar (*sci*) Acting in a single plane, particularly in engineering applied to forces acting in the same plane, i.e. a system of coplanar forces.

copper (*matls*) A metallic element, symbol Cu, relative density 8.93, melting point 1083°C. Copper is a good conductor of heat and electricity and is highly corrosion-resistant. Thin wire and sheet are easily manufactured because copper is very ductile. Consequently, copper is used extensively for electrical wiring and contacts, and for heating equipment. For many applications it is too soft to be used in its pure form, and it is usually used in its alloy forms of either *brass* or *bronze*.

core (*mech*) Part of a mould that is used for forming a hollow void in a *casting*. The core is made in the same shape as the required void, using a moulding sand plus a binder. The sand is rammed into a *core box* to form the shape and then baked in an oven to cure the binder and make it self-supporting. The core is located and supported in the mould in hollows called *core prints*.

core-box (*mech*) Box that acts as a mould for making a *core*, made in two or more parts. The process of making a core is the reverse of making a *mould*. Sand is rammed into the core-box and then the core-box is opened to remove the core.

core plug (*mech*) It is often difficult to remove the cores from a complex casting, such as the cores forming the coolant passages of the cylinder block of an internal combustion engine, and openings have to be left in the wall of the casting to allow the core sand to be removed. *Core plugs* are then inserted to seal the opening after the core sand has been removed. The core plugs also act as safety devices, relieving the pressure, if the coolant freezes and expands in the vicinity of the core hole.

correlation (*maths*) **1** Extent of the dependence of one occurrence on another. See: *cross-correlation* and *autocorrelation*. **2** Correlation is a measure of the amount of association existing between two variables. See: *linear correlation*.

correlation coefficient (*maths*) Dimensionless number between –1 and 1, giving a measure of the linear dependence of one variable on another; e.g. for two variables, x and y, the correlation coefficient is positive if high values of x are associated with high values of y, and negative if high values of x are associated with low values of y.

corrosion (*matls*) Chemical and/or electrolytic attack of a metal. Corrosion adversely affects the mechanical properties of metals through the reduction of the mass of metal present in components and, in some instances, also bringing about changes in the composition and structure of the metal. **1** Dry corrosion. Metals may corrode through direct oxidation when the oxygen in the atmosphere combines directly with metals to form an oxide on the surface. Some metals such as aluminium lead and zinc form an oxide layer that protects the metal underneath from further atmospheric corrosion. **2** Wet corrosion. The *rusting* of ferrous metals is an oxidation process that requires the presence of both water and oxygen to form the familiar red hydroxide film. Once rusting commences the reaction is self-generating and will continue even after the initial supply of moisture and oxygen is removed. Wet corrosion also occurs owing to metals reacting with dissolved acids in rain. An example is the formation of the green carbonate 'patina' on copper-clad roofs. **3** Galvanic (electrolytic) corrosion. Electrolytic attack takes place owing to *electrolysis* between two dissimilar metals, or in one single metal that contains regions of high stress owing to cold-working, in the presence of an electrolyte. It is therefore bad engineering practice to allow two dissimilar metals to be in contact with each other in the presence of any potential electrolyte such as acid rain. See also: *rusting*; *paint*; *electroplating*.

corrosion inhibitor (*matls*) Chemical compounds added to engineering system fluids to reduce *corrosion*. For example, the *antifreeze* solutions added to the engine of a motor vehicle contain a corrosion inhibitor to protect the metal surfaces of the cooling system against corrosion, soften the water to avoid scale formed from *hard water* and control the acidity of the coolant. Corrosion inhibitors are also added to lubricating oils to reduce oil oxidation and protect bearing alloys against corrosion.

cosine (*maths*) See: *trigonometry*.

coulomb (*elec*) SI unit of electric charge, unit symbol C. It is defined as the charge that is transported when a current of one ampere flows for one second. Electric charge has the quantity symbol Q. See also: *ampere*.

Coulomb friction (*sci*) Also called dry friction as it applies to two clean dry surfaces that move or tend to move across one another. See: *friction*.

Coulomb's laws of dry friction (*sci*) These state that: (1) the frictional force always opposes motion; (2) the frictional force is proportional to the normal force between the surfaces; (3) the frictional force is dependent on the character and finish of the materials involved and on the materials themselves; (4) the friction force is independent of the area of contact between the surfaces; (5) provided the sliding speed between the surfaces is relatively low, the frictional force is independent of the sliding velocity.
counterbore (*mech*) A piloted cutter used to increase the diameter of a hole, coaxially with an existing hole in a material, to sufficient depth to accommodate the head of a cheese-head screw or a socket head (cap head) screw so that it will be flush with the metal surface.
countersink (*mech*) A cutter for producing a 90° chamfer on the edge of holes to allow countersunk head screws to seat correctly.
CPA Abb. (*mech*) Critical path analysis. See: *critical path planning*.
crackle test (*mech*) Test for water in oil by heating a sample of the oil and listening for the crackle of the vaporizing water.
crankcase (*mech*) Structure for housing the *crankshaft and bearings* of any reciprocating machine such as an internal combustion engine, a steam engine, or a pump. Often cast as an integral part of the *cylinder block* which contains the cylinder bores and cooling water passages. See also: *sump*.
crankshaft (*mech*) Main drive shaft of any reciprocating machine such as an internal combustion engine. The crankshaft has offset (eccentric) journals, called big-end journals, that allow the linear reciprocating motion to be transferred into the rotary motion of the crankshaft as shown in Fig. C.11. Connecting rods link the reciprocating member (piston) to the big-end journals, and the crankshaft is supported by the main journals about which the shaft rotates.
crank throw (*mech*) **1** In automobile engineering, the crank throw is usually considered to be the distance from the axis of a

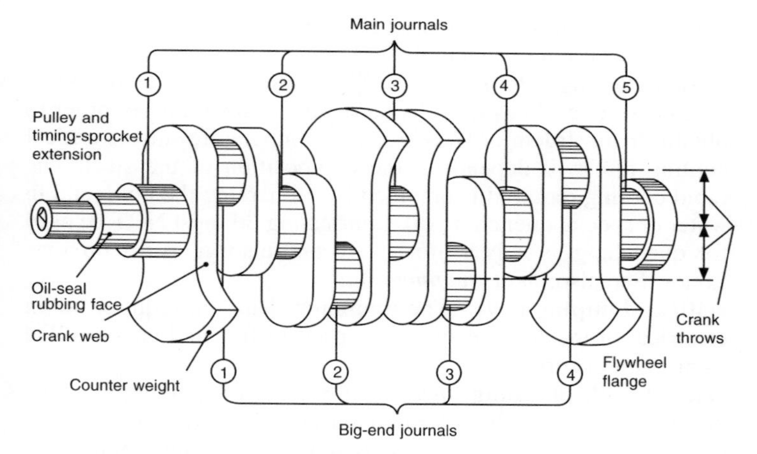

Fig. C.11 Crankshaft Nomenclature

crankshaft to the centre of the big-end journal bearing. Thus the throw is *half the stroke* of the reciprocating member (piston). **2** For general engineering applications the crank throw is usually considered to be *twice the eccentricity* (radius) of the crank and thus equal to the total displacement (stroke) of the reciprocating member.

crank web (*mech*) The webs connecting the big-end journals and main-bearing journals of a crankshaft, as shown in Fig. C.11.

crash (*comp*) The loss of control of a computer program that requires the computer to be reset, e.g. owing to faulty software or incorrect operation.

creep (*matls*) Continuous deformation of a solid metal under a steady load.

crimping (*mech*) **1** Process of folding or bending a metal into ridges to change the shape and/or to provide stiffness. **2** (*elec*) A mechanical method of attaching cable lugs to electric cables when making electrical connections.

critical cooling rate (*matls*) The least rate at which a *steel* must be cooled (quenched) from its hardening temperature in order to obtain a hard structure. The hardening temperature and the critical cooling rate will depend on the composition of the steel. The rapid cooling required is obtained by immersing the hot metal in water or special quenching oil. Lubricating oil must NOT be used for quenching as it is a fire hazard and also gives off noxious fumes. See also: *heat treatment*; *hardening*.

critical damping (*sci*) The minimum damping required in an electrical or mechanical system to prevent free and uncontrolled oscillation occurring.

critical path planning (*mech*) A system for planning a large programme of work in order to determine the optimum sequence of events to ensure the work is completed in the minimum time. Nowadays the procedure is usually computerised and incorporates the sourcing of the materials, components and subassemblies so that they arrive *just in time* to avoid hold-ups and the holding of excessive stocks.

critical point (*sci*) The state of a *fluid* where the liquid and gas phases both have the same density and the fluid is at its *critical temperature*, *critical pressure* and *critical volume*. The two phases cannot easily be distinguished.

critical pressure (*sci*) The pressure of a *fluid* at its *critical point*.

critical speed (*mech*) The rotational speed of a shaft that causes undesirable vibration owing to a periodic disturbing force coinciding with a natural frequency of the machinery. When running up rotating equipment, the acceleration through the critical speed should take place as quickly as possible to avoid damage.

critical temperature The temperature above which a gas cannot be liquefied by an increase in pressure.

critical temperature range (*matls*) The range of temperatures of steel over which the uniform structure of austenite (solid solution of carbon in iron) present at high temperatures changes to ferrite, pearlite and cementite at lower temperatures. The lower limit is approxi-mately 723°C and the upper limit varies according to the carbon contents, as shown in Fig. C.12. When a steel contains less than 0.83%, carbon grains of ferrite and pearlite are produced below the lower critical temperature. (Pearlite grains have a lamellar structure of alternate layers of ferrite and cementite.) For steels with a carbon content above 0.83%, pearlite grains are produced

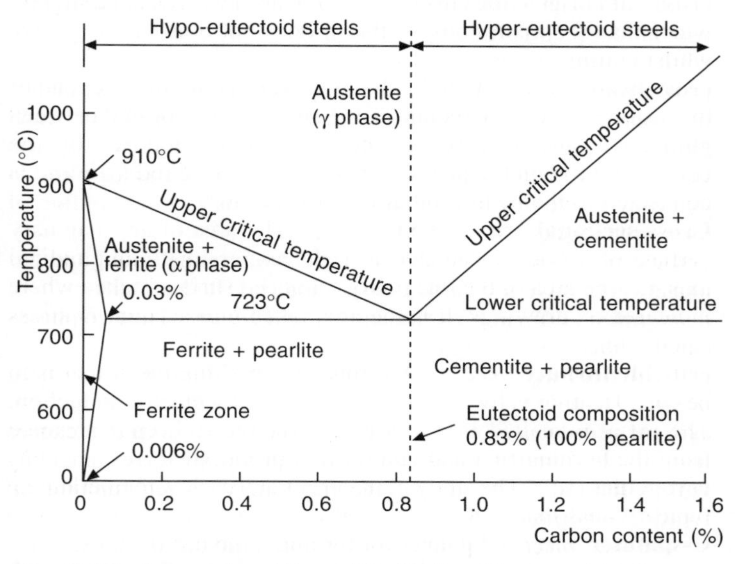

Fig. C.12 Iron-carbon Phase Equilibrium Diagram (steel section) for Critical Temperatures in the Solid State

and the excess cementite (iron carbide) is deposited at the grain boundaries at temperatures below the lower critical temperature. When a steel contains exactly 0.83% carbon (the eutectoid composition), the composition of the steel changes directly from uniform austenite to pearlite at 723°C. The large range of properties of steel can be achieved by altering the carbon content and the heat-treatment process.

critical volume (*sci*) The volume of unit mass of a fluid at its *critical point.*

cross-correlation (*elec*) The multiplication of two signals that are averaged over a time interval.

cross-cut chisel (cape chisel) (*mech*) Chisel with an edge slightly wider than the main body so that it does not bind in the groove whilst cutting.

cross-head (*mech*) A sliding bearing attached to the outer end of the piston rod of some reciprocating engines. It is located between guide bars and absorbs any bending forces imposed by the connecting rod on the piston rod. The connecting rod *little end* is connected to the piston rod at the cross-head.

Cross-sectional area (CSA) (*mech*) The area of an imaginary surface of a body calculated at right angles to its longitudinal axis, i.e. the area of the imaginary 'chopped surface' taken where indicated on drawings. It is the area taken into account in stress calculations.

crucible furnace (*mech*) A furnace for melting metal that is to be cast. Heating is by gas or oil burners, or by electric induction. The metal is melted in a refractory crucible to keep it separate from the burning fuel and combustion products.

cryogenics (*sci*) The study of materials at very low temperatures, typically less than 1 K.

C-spanner (*mech*) Spanner for turning large narrow nuts. It has a head in the shape of a 'C' with a projection that fits into a notch in the nut.

cumulative frequency distribution curve (*stats*) A graph produced by plotting cumulative frequency against upper class boundary values. The coordinates are connected by straight lines or a curve. The cumulative frequency axis is vertical and the class boundary axis is horizontal. The curve so produced is often referred to as an ogive curve. See: *ogive curve.*
cupola furnace (*matls*) A small *blast furnace* for melting and refining *pig-iron* in order to produce castings in an iron foundry. The furnace is lined with firebrick and the charge is made up of pig-iron, steel scrap, coke and limestone added in layers. Combustion air (the *blast*) is then blown into the furnace towards its bottom, just above the tapping holes.
current (*elec*) The rate at which electrons or electrical charge in an electrical circuit move in a particular direction. Electrical current is measured in amperes often abbreviated to 'amps', quantity symbol I, unit symbol A. A current of one ampere is equal to a charge flow of one *coulomb* per second.
current density (*elec*) The current flow through a conductor per unit cross-sectional area, measured in amperes per square metre.
current gain (*elec*) The ratio of the output current of an amplifier to the input current of an amplifier, i.e. gain = $(I_{out}) \div (I_{in})$.
cursor (*comp*) A flashing character, usually a horizontal or a vertical short line, on a computer VDU, to indicate the current display location.
cut and paste (*comp*) A method of transferring data, text or a diagram between different sections of a document, between documents or between different software, via the *clipboard.*
cut-off frequency (*elec*) See: *bandwidth.*
cutter diameter compensation (*CAE*) The programmed cutter path follows the axis of the cutter and the controller automatically makes allowance for the diameter of the actual cutter being used,

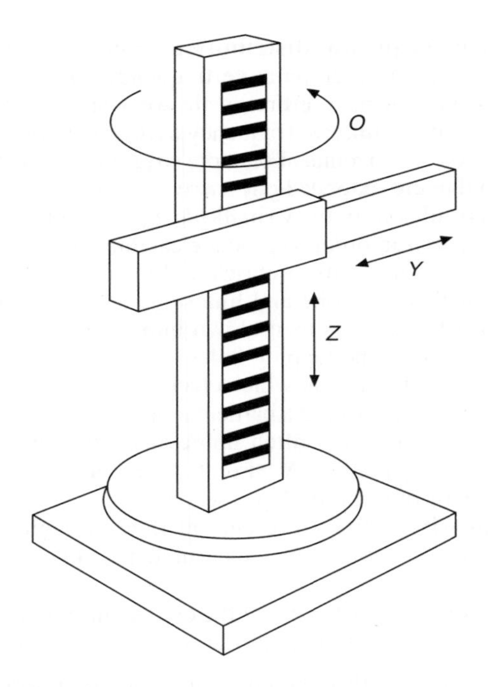

Fig. C.13 Cylindrical Coordinate Robot

thus simplifying programming. Again this allows for changes in cutter diameter owing to regrinding without having to rewrite the programme.

cybernetics (*comp*) Study of control and communication systems in living organisms and complex electronic circuits.

cyberspace (*comp*) See: *virtual reality*.

cylindrical coordinate robot (*CAE*) These have two orthogonal and one rotary axis, as shown in Fig. C.13. The horizontal arm telescopes in and out and moves vertically up and down the column which rotates on the base. Therefore the working envelope (volume) is cylindrical. The resolution of a cylindrical robot, using a standard rotary encoder, is not constant but depends on the extension of the arm. With the arm extended one metre from the column axis, the positional resolution of the wrist assembly is of the order of 3 mm compared with a resolution of a constant 0.01 mm for a Cartesian robot.

D

Dalton's law of partial pressures (*sci*) The pressure of a mixture of gases is equal to the sums of the pressures of the individual gases if each were to occupy the same volume as the mixture at the temperature of the mixture.

damping (*sci*) A reduction of the *amplitude* of a dynamic oscillatory system. In mechanical systems this is mainly due to frictional resistances draining energy from the oscillatory system. The amount of damping of a dynamic system is measured by the damping ratio, usually represented by the Greek letter ζ. An oscillatory system with a damping ratio of 0 has zero damping and produces *simple harmonic motion*. As the value of ζ increases towards 1 the tendency of the system to oscillate reduces; when $0 < \zeta < 1$, the system is said to be *underdamped*. A system with a damping ratio of 1 will have no oscillatory component, and is said to be critically damped; a critically damped system comes to rest in the shortest possible time. If the value of ζ increases above 1 the system is said to be *overdamped* and becomes increasingly sluggish as the response time increases.

database (*comp*) System for storing data, usually computerized, with information classified in different categories for extraction purposes. For example, consider a company database that stores the names, addresses and telephone numbers of customers. Information about each customer is stored in a single record. Each record is divided up into a series of fields, in this case it could be four fields: surname, forename, address and telephone number. The database can then assist in comprehensive search procedures, such as extracting the details of anyone with a forename of, say, John or anyone living in a particular town.

Datum (*mech*) A line, edge or surface used as a basis for measurement. (1) *Line datum.* This can be the centre line of a

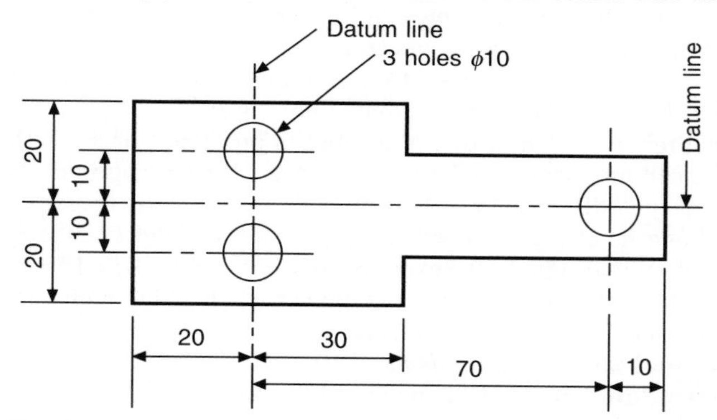

(a) Line datum

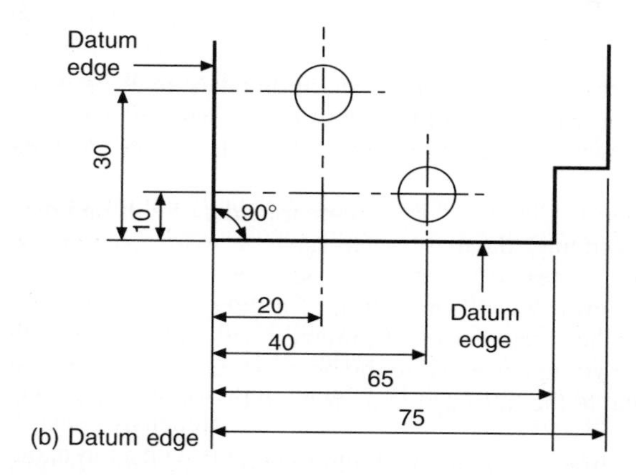

(b) Datum edge

Fig. D.1 Datum Line and Edge

symmetrical component as shown in Fig. D.1(a). (2) *Edge datum.* An edge or, more usually, a pair of mutually perpendicular edges from which measurements can be taken as shown in Fig. D.1(b). (3) *Surface datum.* See: *marking out table.*

de-aerator (*mech*) Vessel for removing dissolved air from boiler feed water, by heating at reduced pressure.

decibel (*elec*) unit dB. A decibel is one tenth of a *bel*. It is a logarithmic measure of the ratio of the relative amplitude of two signals: $n\,(\text{dB}) = 20 \log_{10} V_1/V_2$, where V_1 and V_2 are the two signal voltage levels; when used for expressing the ratio of the power levels of two signals, the expression becomes: $n\,(\text{dB}) = 10 \log_{10} P_1/P_2$ where P_1 and P_2 are the two power levels being compared.

deciles (*stats*) See: *quantiles.*

decimal (*maths*) See: *denary.*

dedendum (*mech*) Distance from the pitch circle of a gear to the root circle.

dedicated computer (*comp*) A computer that is permanently assigned to one operation only, as in the controller of a *CNC* machine tool.

definite integral (*maths*) When integration takes place between limits, the unknown constant of integration (C) cancels out and a numerical answer can be calculated. Hence the name 'definite integral'.

delimiter (*comp*) Character used to separate data fields and data records in a computer database, or to separate words, sentences, paragraphs and pages in a word-processing program.

delta connection (*elec*) Connection of a three-phase electrical system where the three phases of voltage form a closed triangle so that the line voltage is equal to the phase voltage and the phase current is equal to the line current multiplied by $\sqrt{3}$. See also: *star connection.*

de Morgan's theorem (*maths*) Technique of *Booloean algebra* used to simplify logic circuits, usually into NAND or NOR functions for ease of manufacture. Basically, this theorem states that the negative of the sum of two classes is equal to the product of their negatives, and the negative of their product equals the sum of their negatives; i.e. to change the sign of an expression such as (A + B), negate (make negative) each variable and change each AND to OR, each OR to AND. For example:

$$\overline{A + B} = \overline{A} \cdot \overline{B}$$

and

$$\overline{A \cdot B} = \overline{A} + \overline{B}$$

demultiplexer (*elec*) Computer circuit that allows data from a single source to transfer to one of several destinations. See: *multiplexer*.

denary (*maths*) The system of numbers in everyday use comprising the 10 digits from 0 to 9 inclusive, i.e. it has a base or radix of 10. Thus the denary number 3784 is equivalent to $3 \times 10^3 + 7 \times 10^2 + 8 \times 10^1 + 4 \times 10^0$. Thus it is the sum of the terms where each term is a digit multiplied by the base raised to some power.

denominator (*maths*) See: *fraction*.

density (*sci*) The mass of a unit volume of a substance, measured in kg/m^3 in the SI system. The quantity symbol commonly used in calculations is the Greek letter ρ, and is calculated as follows:

$$\rho = \frac{\text{mass of a material, } m}{\text{volume occupied, } V}$$

As volume varies with temperature, the temperature at which the density of a substance was determined should also be stated.

dependent event (*stats*) An event for which the probability of the event occurring is affected by the probability of another event occurring.

determinants (*maths*) The coefficients forming the array within a matrix. See: *array*; *matrix*.

deterministic signal (*elec*) Signal that can be described completely in a precise way at all instants in time, e.g. the sinusoidal electrical current supplied from the mains. Deterministic signals can be *periodic* or *aperiodic*.

dew point (*sci*) The temperature at which the vapour in the air is saturated and will deposit droplets of water.

dezincification (*matls*) When ordinary brass alloys containing only copper and zine are subjected to a marine environment the zine is gradually dissolved out of the alloy leaving only a weak spongy mass of coper. This is called *dezincification*. To prevent dezincification happening, brass alloys intended for use in a marine environment contain a trace of tin. See: *admiralty brass* and *naval brass*.

dial test indicator (DTI) (*mech*) Also referred to in engineering workshops as a 'clock' because of its appearance. The dial test indicator is a measuring instrument in which the displacement of a plunger or a stylus is magnified mechanically and the magnitude of the deflection is indicated by the movement of a pointer over a suitably calibrated dial. Nowadays electronic instruments giving a digital read-out are increasingly employed. They are used extensively in workshops for testing parallelism, squareness, roundness and concentricity, and also for machine setting.

diaphragm pump (*mech*) A reciprocating pump in which the conventional piston or cylinder is replaced by a flexible diaphragm and non-return valves. The diaphragm is sealed around the edges and the pumping motion is activated by a mechanism that causes the centre of the diaphragm to rise and fall.

die (*mech*) **1** General term for many forming and cutting tools in engineering. For example, wire is drawn through a die to reduce its diameter; shapes are stamped out of sheet metal using a punch and die combination; coins are made by pressing the metal discs into an impression cut in a die. **2** Screw threads, split dies and button dies held in a die-stock are used for producing external screw threads.

die-casting (*mech*) Casting carried out by injecting molten metal under high pressure into a hollow metal mould (die). Very high rates of production can be achieved and the surface finish and accuracy of the casting is sufficiently good to avoid the necessity for many machining operations. The process is limited to zinc and aluminium alloys having a low melting temperature. Higher melting point alloys such as aluminium bronze can be gravity die-cast in a similar way to conventional casting but using metal dies instead of sand moulds.
dielectric (*elec*) An electrical insulator to which an applied electrical field causes an electric charge displacement but no flow of charge, other than the initial charging current. See also: *capacitance*; *permittivity*.
dielectric breakdown (*elec*) Forced conduction across an insulating material when a maximum working voltage (*dielectric strength*) is exceeded; the resulting short circuit and resulting surge of current causes the destruction of a capacitor.
dielectric constant (*elec*) See: *permittivity*.
dielectric strength (*elec*) The electric stress necessary to break down a dielectric; the voltage at which *dielectric breakdown* occurs and insulation ceases, expressed in units of kV/mm thickness. The electric stress, steady or alternating, is usually applied for one minute when testing.
die-stock (*mech*) A hand tool for holding threading dies with two torque arms (handles) for rotating the die and screws for opening or closing the die in order to adjust the die diameter.
differential amplifier (*elec*) Common electronic circuit used to amplify the difference between two input voltages. The output depends only on the *difference* between the two voltages and is independent of the absolute signal levels.
differential calculus (*maths*) A branch of mathematics dealing with continuously variable quantities based on the differential coefficient, or derivative, of one quantity with respect to another

of which it is a function. Differential calculus is the inverse of integration and can be used to determine the gradient of a curve at any given point. For example, if $y = ax^n$ then $\mathrm{d}y/\mathrm{d}x = anx^{(n-1)}$, where $\mathrm{d}y/\mathrm{d}x$ is the differential coefficient of $y = ax^n$.

differential equations (*maths*) These are equations containing differential coefficients. They are classified according to the highest derivative that occurs within them. For example:

$$\frac{\mathrm{d}y}{\mathrm{d}x} = 7x$$

is an equation of the first order, whilst

$$\frac{\mathrm{d}^2 y}{\mathrm{d}x^2} + 5\frac{\mathrm{d}y}{\mathrm{d}x} + 2y = 0$$

is an equation of the second order. Differential equations are widely used in engineering and science. There are several different types of differential equation and each type requires its own method of solution. See also: *Laplace transforms* (Appendix 1).

digital (*elec*) A digital signal has only a finite number of values, unlike an *analogue* signal which can change value by infinitely small amount. Most digital signals are of *binary* form in which there are only two possible states, as used by digital computers and other digital devices.

digital-to-analogue converter (*comp*) (DAC) A device for transforming a digital signal inside a computer to a proportional *analogue* signal. The smallest possible change in the output is called the resolution and this can be calculated for an N-bit converter as follows:

$$\text{resolution} = \frac{100}{2^N}\%$$

e.g. an 8-bit converter has a resolution of

$$\frac{100}{2^8}\% = 0.39\%$$

The signal also changes from a *discrete* signal to a *continuous* signal. Computer control requires a digital-to-analogue converter to transform the calculated digital control effort into a voltage or a current signal that can be applied to the actuator of the control system.

dimensional word (*CAE*) These are words related to a linear dimension when writing a CNC part program, i.e. any word commencing with the letters X, Y, Z, I, J, K or any word in which the above characters are implied. The letters X, Y, Z, refer to machine axes and the letters I, J, K, refer to circles and arcs of circles. Older systems did not use the decimal point but used leading and trailing zeros to indicate the position of the decimal point, e.g. 35.6 mm would be written 0035600. In modern systems the dimension is entered directly as 35.6. In addition to the dimension, the axis also has to be stipulated, e.g. if the movement is along the x-axis then the word would be written X35.6. Finally, the direction of movement has to be added. Absence of a sign indicates positive movement and a minus sign in front of the word indicates a negative movement. For the z-axis a minus sign indicates that the tool moves towards the cutting tool. This is a safety feature since omission of the sign moves the tool away from the work. See: *management word.*

diode (*elec*) A two-terminal electronic device that may be of thermionic or solid state construction. Nowadays diodes are almost entirely solid state (p–n junction devices). They are unidirectional devices and fall into two main categories: rectifier (power) diodes and signal diodes. The former have to handle large currents and sustain high peak inverse voltages. They are used to convert alternating currents into direct currents and are only called upon to operate at relatively low frequencies. The latter, signal diodes, are used as logic elements or as demodulators in r.f. circuits. The currents and voltages are small, but the devices have to operate at very high speeds. All diodes have a non-linear relationship between the forward current and the voltage drop. A diode can be used as a one-way conductor as the reverse current is small compared to the forward current until the reverse breakdown voltage is reached (see: *avalanche current*). The exception to this is the *zener diode* designed to operate in a circuit with a reverse current to provide a constant voltage equal to its breakdown voltage. In this way it is used as a reference voltage in voltage stabilizer circuits.

direct current (d.c.) (*elec*) An electric current that involves a flow of charge in one direction only, as created by a battery or from the electrical mains after rectification. See also: *alternating current*.

direction of flow (*elec*) (1) *Traditional current flow*: this adopts the historical convention of electric current flow from positive to negative; many of the rules governing electromagnetism and electromagnetic induction still use this convention. (2) *Electron current flow*: atomic research now shows that most electric currents consist of electrons flowing from negative to positive, since negatively charged electrons are attracted to the positive pole of a system. Electronic engineering theory uses electron current flow.

direct memory access (DMA) (*comp*) Method of transferring data directly to and from a computer memory using a hardware device called a direct memory access controller, rather than by program instructions; this results in a much faster data transfer.

direct numerical control (*CAE*) In direct numerical control (DNC) the program is prepared on a remote computer using simulation software that shows the shape of the program on the visual display unit (VDU) and also carries out all the calculations. The program can be checked and edited as required. Hazards such as collisions with clamping devices can be eliminated. Finally, post-processing software converts the computer software language into a language suitable for loading into the CNC machine's controller. This is very convenient when a range of machines and controllers are in use, since the programmer only has to learn the one language as used on the computer. The program can be transferred to the machine on disk or downloaded directly to the machine from the computer. See: *manual data input*; *punched tape input*; *conversational input*; *magnetic tape input.*

directory (*comp*) List of subdirectories, or programs or data files, i.e. part of a hierarchical file structure of microcomputer operating systems. The root directory of the hard disk contains the names of subdirectories and the names of files. Each subdirectory can refer to files or to further subdirectories.

direct strain (*matls*) See: *strain.*

direct stress (*matls*) See: *stress.*

discrete 1 (*elec*) A signal that is defined only at specific instants in time, usually as a result of a continuous analogue signal being sampled at equally separated time intervals. In computer control systems the sampling is carried out by a computer, and so the discrete signal will also be digital. **2** (*stats*) Data that are obtained by counting, where only whole numbers are possible.

disk formatting (*comp*) Process of preparing a magnetic computer disk before use by establishing control data, such as track and sector number.

displacement (*sci*) Vector quantity (s) that describes the distance and direction when a body moves from one point to another, measured in metres in the SI system.

distributed load (*sci*) Assumption made in mechanical load problems, such as the bending of beams, to simplify calculations, that forces or loads are spread uniformly over the length of a beam, usually to account for the weight of the beam itself.

disturbance inputs (*sci*) Inputs to a plant other than the controlled inputs that can affect the dynamics of the system.

division (*maths*) The mathematical operation of determining how many times one number (the *divisor*) is contained within another number (the *dividend*), the result is called the *quotient*. If the divisor is not contained an integral number of times in the dividend, the number left over is called the remainder.

dolly (*mech*) Tool for supporting the head of a rivet whilst forming a head on the other end using a rivet 'snap'.
doping (*elec*) The controlled addition of trivalent or pentavalent materials (impurities) to high-purity, *intrinsic* semiconductor materials such as silicon and germanium in order to convert them into *extrinsic* semiconductor materials and thus to alter the type of dominant conduction (*n*-type or *p*-type).
DOS Abb. (*comp*) Disk operating system, a microcomputer operating system.
DOS shell (*comp*) A method of temporarily suspending the operation of a program to access the DOS operating system.
dowel (*mech*) A hardened and ground cylindrical steel pin with a short taper lead (chamfer) that can be used to provide a positive location between two plates or components. The dowel should be a light drive fit into corresponding location holes accurately machined in both of the parts to be joined.
drag (*mech*) Bottom half of a sand moulding box used for casting. See: *cope*.
dresser (*mech*) Any workshop device used for dressing and truing grinding wheels. Dressing is the process of removing blunt grains of abrasive and/or particles of metal loading the spaces between the grains. Truing is the process of cutting the surface of the rotating abrasive wheel with a diamond so that the periphery of the wheel runs true with the spindle axis of the grinding machine. Examples of dressing and truing devices are shown in Fig. D.2.
drift (*mech*) **1** Tool for finishing-off and cleaning up small non-circular holes, e.g. square, hexagonal, etc.; a drift matching the required hole dimensions is hammered or pressed through the hole. The drift may have teeth and is hardened and tempered. **2** Tool for driving out a key or pin from a shaft. **3** A wedge-shaped tool used for removing taper shank drills from drilling machine spindles.
drill (*mech*) See: *twist drill*; *drilling machine*.
drilling machine (*mech*) A machine for rotating a drill bit and

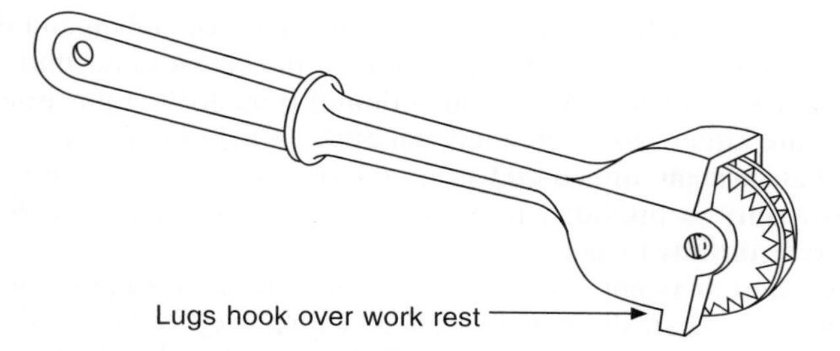

(a) Huntington wheel dresser for dressing the grinding wheel of an off-hand grinding machine

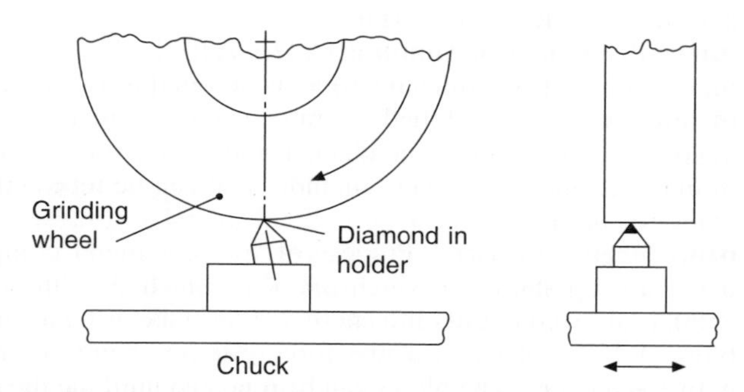

(b) Dressing and truing the grinding wheel of a precision grinding machine using a diamond

Fig. D.2 Grinding Wheel Dressing Tools

for providing the axial force necessary for feeding the drill into the material being drilled. Typical drilling machines are shown in Fig. D.3. **1** Bench drilling machine. A small machine for mounting on a workbench having a sensitive hand-feed mechanism and the facility to change the spindle speed by means of cone pulleys (see: *sensitive feed*). **2** Pillar- and column-type drilling machines. These are larger, floor-mounted machines. A gear boxor variable-speed electric motor is used to vary the spindle speed and a power feed facility is provided in addition to manual feed. **3** Radial-arm–type drilling machines. These machines are used on the largest work where it is not convenient to move the work to position it under the drill. An all-geared drill head with power feed is mounted on an arm that in turn can be swung about a vertical column. The radial arm can be raised or lowered on the column and the drill head can be traversed to or from the column along the arm. Thus the drill can be easily positioned over the work.

driver (*comp*) Software program for controlling a computer peripheral device, e.g. a printer.

'drive-through' programming (*CAE*) A method of programming industrial robots in which the robot is driven through its sequence of movements by a skilled operator using a remote, hand-held keypad called a teaching pendant. The motion pattern is recorded in the controller's computer memory and can be repeated as and when required.

drop forging (*mech*) Process of forming metal components using a drop stamp in which one die, attached to the hammer (tup), is allowed to fall onto another (fixed) die using a controlled blow. A billet of metal at the forging temperature is entrapped between the dies. The blows can be repeated until the die cavities are filled. This process is used for producing a large number of identical components of high strength and reliability.

drop stamp (*mech*) Device for producing drop-forgings using closed dies. It consists of an anvil that supports the bottom die and a hammer (tup) that carries the top die. The side frames guide the hammer and support the mechanism for raising the hammer and allowing it to fall in a controlled manner. The hammer is

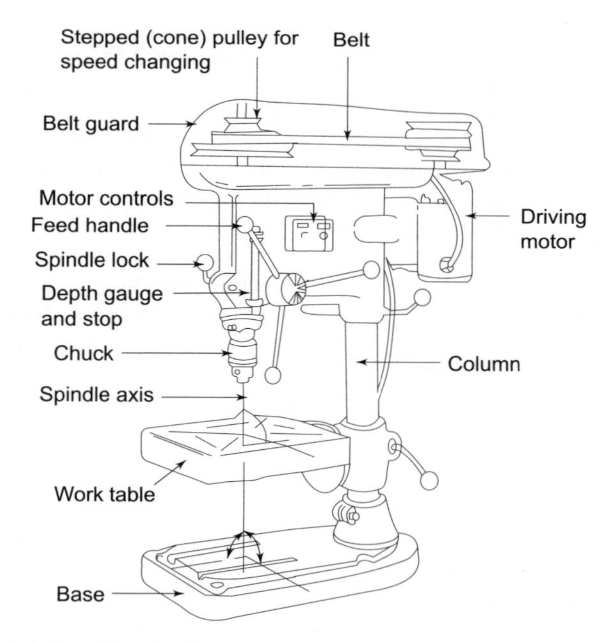

Fig. D.3 (a) Bench drilling machine

lifted by a leather belt (*belt* or *strap hammer*) or by a flat wooden board (*board hammer*). The belt or the board are trapped between the power roller and a pinch roller when the hammer is to be raised. The capacity of a drop stamp is expressed in terms of the mass of the top die block and the force of the blow. Drop stamps are nowadays being replaced by forging presses which are more easily automated.

dryness fraction (*sci*) A measure of the quantity of *saturated vapour* (usually steam) present in a unit mass of wet vapour, with

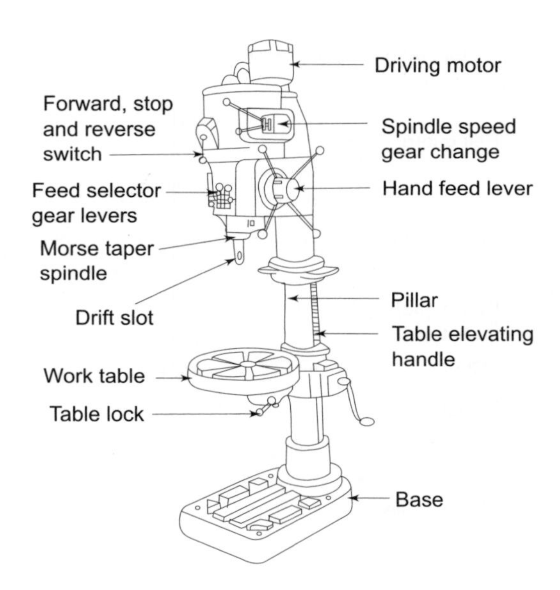

Fig. D.3 (b) Pillar type drilling machine

values between 0 and 1. A *saturated fluid* has a dryness fraction of 0, whereas a saturated vapour which is completely dry has a dryness fraction of 1.

dry steam (*sci*) Steam that is a pure gas with no liquid present, i.e. steam at or above the saturated vapour temperature at the given pressure. Dry steam usually refers to *saturated vapour* and steam above this point (unsaturated vapour) is referred to as *superheated steam*.

DTL (*elec*) Diode transistor logic, an early form of integrated circuit technology.

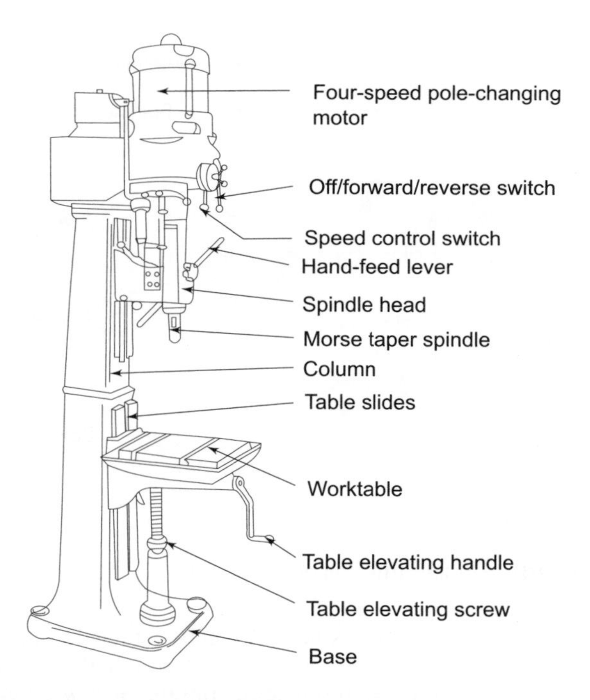

Fig. D.3 (c) Column type drilling machine

ductility (*matls*) The ability of a metal to be formed by plastic flow without breaking when subjected to a tensile force, i.e. forming by stretching. Material for making wire needs to be ductile so that it can be drawn through a die without breaking. Not to be confused with *malleability*.
duralumin (*matls*) Trade name of a commonly used heat-treatable, wrought, aluminium alloy, containing 4% copper plus traces of

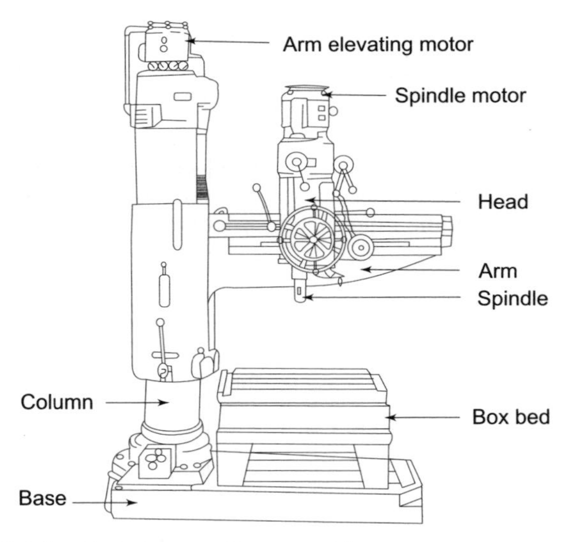

Fig. D.3 (d) Radial-arm drilling machine

manganese, magnesium and silicon. It can be softened by annealing at 480°C (see: *solution treatment*) after which it age-hardens naturally over 4 days at room temperature. See: *precipitation treatment*.

dynamic imbalance (*mech*) Out-of-balance forces that can occur when a body (usually a wheel or rotor) rotates. These out-of-balance forces cause vibrations that can be seriously damaging and result from rotating masses that do not act in the same plane. Dynamic imbalance can occur in a system that is balanced in the static condition.

dynamic system (*sci*) A system whose output response is a function of both the inputs and time. The output of most engineering systems changes with time and corresponding mathematical models include time as a variable.

dynamo (*elec*) A d.c. generator in which the field magnet is stationary and the armature conductors rotate. The alternating current induced in the armature conductors is rectified by a commutator and brush gear system. The magnitude of the output voltage can be controlled by varying the strength of the current flowing through the field-magnet windings.

E

earth (*elec*) The connection to Earth, by a conductor of low resistance, of one pole or the neutral point of an electrical supply system for safety reasons. The connection to earth of any metal component or casing not in the electrical circuit.
earth fault (*elec*) Electrical fault that causes a connection between the voltage source and earth. See also: *earth leakage protection system.*
earth leakage protection system (*elec*) Fault protection device for an electrical system, comprising a circuit-breaker that trips when an earth fault occurs, thus disconnecting the supply from the faulty parts quickly. The system operates by detecting imbalances of currents in the live and neutral conductors. See: *earth fault.*
eccentric (*mech*) A device for converting rotary motion into linear motion. Unlike a crank and connecting rod, it cannot work in the opposite direction and convert linear motion into rotary motion. An eccentric is more robust and compact than a crank and connecting rod but more limited in stroke. Often used for

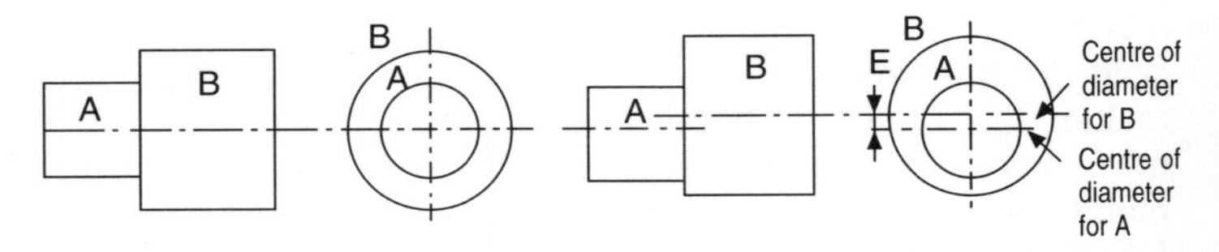

Fig. E.1 Concentricity and Eccentricity (a) concentric diameters – both have the same centre; (b) eccentric diameters – each diameter has a different centre

driving reciprocating pumps and in forging and coining presses.
eccentricity (*mech*) Displacement of one axis relative to another in the same component, as shown in Fig. E.1. See: *concentricity*.
ECL (*elec*) Emitter coupled transistor logic; contemporary integrated circuit technology with typical characteristics of: fan-in = 5, fan-out = 50, propagation delay = 1.1 ns, power consumption = 30 mW. This technology finds applications in computers where its high-speed characteristics are essential.
economizer (*mech*) Device fitted in the flue of a steam boiler for recovery of some of the heat energy from the flue gases that would otherwise be wasted, thus increasing the plant efficiency. In a common steam power cycle the boiler supplies steam to power a turbine for driving an electrical generator; the exhaust steam is then condensed and returned to the boiler by a feed pump. Feed water may pass through an economizer, which is comprised of a bank of tubes situated in the boiler flue. This raises the water temperature prior to it entering the boiler. Care must be taken not to drop the temperature of the combustion gases below the dew point or corrosive acids may be formed in the flue.
eddy current (*elec*) The current induced in any electrically conductive material when linked with a changing magnetic field, particularly in the core of a solenoid; this causes a heating effect, and represents an energy loss to transformers and electrical machines which involve an alternating current and a subsequent alternating field. Eddy current losses are minimized in most equipment by making the core out of thin iron laminations insulated from each other, thus increasing the electrical resistance of the core whilst maintaining a low magnetic *reluctance*.
EEPROM Abb. (*comp*) Electrically erasable programmable read-only memory. A computer read-only memory which can be programmed by the user and erased by an electrical signal.
efficiency (*sci*) A measure of the performance of a system,

calculated from the ratio of the net energy or power delivered by the system to the net energy or power fed into the system, usually expressed as a percentage. The thermal efficiency of a heat engine is the ratio of the latent energy in the fuel supplied to the useful mechanical work delivered.

elastic collision (*sci*) Collision that occurs between two elastic bodies involving zero loss of energy through sound, heat, etc., so that the total kinetic energy before the collision equals the total kinetic energy after the collision; not possible in reality as some kinetic energy is always converted into other forms of energy.

elasticity (*sci*) Ability of a material to return to its original dimensions after an applied load has been removed.

elastic limit (*sci*) The limit to which a material may be stretched by an applied load before a permanent change in the original dimensions takes place. Below the elastic limit the material will return to its original size and shape when the applied load is removed. Above this limit, permanent plastic deformation takes place. See: *limit of proportionality*; *modulus of elasticity*; *Hooke's law*.

elastic sensing element (*sci*) Sensing element of an instrumentation system that converts the quantity to be measured into a readable signal by allowing a change in force or pressure to produce a corresponding change in the element's shape or length.

elastomer (*matls*) A natural or synthetic material (e.g. a rubber) that has a very high *elasticity*.

electric arc welding (*mech*) A fusion welding process in which the heat energy to melt the edges of the parent metal is derived from an electric arc struck between the workpiece and an electrode. The electrode is also melted and acts as a filler rod. It is coated with a flux to reduce oxidation and produce a protective shield of carbon dioxide and a protective layer of fusible slag over the weld.

electric current (*elec*) See: *current*.

electric field strength (*elec*) Force per unit charge (F/Q) exerted on a particle in an electric field, quantity symbol E, measured in units of newtons per coulomb (N/C).
electric flux (*elec*) Imaginary lines of electric force that represent an electric field existing between two charged bodies, quantity symbol ψ (Greek letter psi), measured in units of coulombs (C), which has the same numerical value as the charge that produces it.
electric flux density (*elec*) The *electric flux* per unit area, symbol D, measured in units of coulombs per square metre (C/m^2).
electric furnace (*matls*) Any furnace that uses electricity as its energy source. Although more expensive than gas or oil, electricity is cleaner and more easily adapted to automatic control. In addition there are no products of combustion to contaminate the material being heated. **1** *Electric arc furnaces* are used for the bulk production of high-quality alloy steels, the arc being struck between carbon electrodes and the molten metal bath. Impurities are oxidized by a covering of selected slag. **2** *Electric induction furnaces* use a high-frequency alternating current source to induce eddy currents in the furnace charge itself as the heat source. Where very high purity is required the metal is melted under high vacuum conditions. **3** *Electric resistance furnaces* are used for lower temperature processes such as heat treatment. The furnace is heated by means of electric resistance elements surrounding the furnace charge. The principle is much the same as for a domestic electric oven.
electricity (*elec*) An energy source associated with static or dynamic electric charges or associated with the migration of the electric charge of protons and electrons in a material.
electric motor (*elec*) Machine having a *rotor* and a *stator* for converting electrical energy into mechanical energy; the driving force being produced by an interaction of the magnetic field of a current-carrying coil or armature in a static magnetic field. See also: *synchronous machine*; *induction motor*.

electrode (*elec*) A conductive rod that collects or delivers electric charge. See: *anode*; *cathode*.

elctrolysis (*sci*) Electrochemical reaction that takes place when an electric current is passed through an *electrolyte* between two electrodes, an anode bearing a positive charge and a cathode bearing a negative charge. The ions in the electrolyte act as electron carriers. Negative ions migrate to the anode which attracts electrons from the electrolyte. Positive ions migrate to the cathode which emits electrons to form neutral atoms (see also: *Faraday's laws of electrolysis*; *electroplating*).

electrolyte (*elec*) An liquid containing positive and negative ions that may conduct electricity. It may be an aqueous solution or a molten solid.

electrolytic corrosion [*matls*] Also called galvanic corrosion. A corrosion mechanism occurring when two dissimilar metals in contact with an electrolyte (such as acid rain) form an electrolytic cell. This causes an electric current to circulate within the system. The metal that is more electronegative in the electrochemical series will be eaten away, whilst the metal that is more electro-positive in the electrochemical series will be protected. Thus, in the case of galvanized iron, the zinc is slowly eaten away whilst it protects the iron (the zinc is *sacrificial*). In the case of tin plate, the tin is more electropositive than the iron so the iron tends to be eaten away. For this reason care must be taken to seal any cut edges with soft solder when working with tin plate. See: *electrolysis*.

electromagnet (*elec*) A device consisting of a current-carrying coil (solenoid) wound around a soft-iron core in order to produce a magnetic field. Unlike permanent magnets, the magnetic field produced may be varied in strength by controlling the current through the conducting coil, reversed by changing the current direction, and turned on and off. Electromagnets have many engineering applications including *solenoid valves and relays*.

electromagnetic brake (*mech*) Machine brake that produces the braking force by pressing together two friction surfaces by means of an *electromagnet*.
electromagnetic clutch (*mech*) *Friction clutch* operated by means of an *electromagnet* rather than springs.
electromagnetic generator (*elec*) Rotary machine that produces electrical energy from mechanical energy due to *electromagnetic induction*; the drive for the generator is some form of engine such as a diesel engine, a steam turbine, or a water turbine, for example.
electromagnetic induction (*elec*) The production of an *e.m.f.* in a conductor (usually a coil) when there is relative movement between the conductor and a magnetic field with which it is linked. This can be due to the magnetic field changing as created by an alternating current or to the conductor moving through a constant magnetic field. The value of the induced e.m.f. is proportional to the rate of cutting of the magnetic *flux*, and the number of turns on the conductor coil. Expressed mathematically,

$$\text{e.m.f.} = -N\frac{\mathrm{d}\Phi}{\mathrm{d}t}$$

where: N is the number of turns of the conductor coil, Φ is the magnetic flux, and $\mathrm{d}\Phi/\mathrm{d}t$ is the rate of change of flux. The expression above is often described as the 'rate of change of flux linkage'. The negative sign in front of the expression indicates that the polarity of the induced e.m.f. always opposes the change that produced it, in accordance with *Lenz's Law*. See also: *Faraday's laws*; *inductance*.
electromotive force (e.m.f.) (*elec*) The driving force in an electric circuit that causes free electrons in a conductor to drift in a given direction; symbol E, measure in volts (V). This should not be confused with potential difference, which is a difference in voltage levels as current flows through a resistive circuit. The e.m.f. causes

the current to flow whilst the potential difference is a result of current flow.

electron beam welding (*mech*) Method of joining metals together in which a high-speed stream of electrons is sharply focused and used to produce intense local heat. The welding process is carried out in a vacuum to prevent the beam from dispersing on hitting air molecules. It is a useful process for welding metals having a high melting temperature and also for welding reactive metals.

electronic mail (e-mail) (*comp*) A mail system for computer networks. A user on one system can send a message to a user on the same or a different system. The mail is sent to an electronic 'mail box' on a computer hard disk. The user can then periodically check the mailbox for mail messages. Messages can then be read or printed and then discarded. The system is fast, convenient and cheap.

electroplating (*matls*) A process for coating a surface, by electrolytic action, with a thin layer of metal which may be protective, decorative or both. The components to be plated are immersed in an *electrolyte* and are made the *cathode* of an electrolytic cell. To complete the circuit, anodes are also immersed in the electrolyte. The work is connected to the negative pole of a d.c. source and the anode is connected to the positive pole of a d.c. source. Upon the passage of a current through the cell, ions of the coating metal are transferred from the electrolyte to the surface of the work where they are deposited as metal atoms to build up the required coating. The greater the current density and the longer the process time taken, the greater will be the thickness of the plated coating. In most electroplating processes the strength of the electrolyte is maintained by dissolving metal ions from the anode. In some processes such as chromium plating the anode is unaffected and the strength of the electrolyte has to be maintained by replenishment.

electrostatic generators (*elec*) These use the friction between suitable materials that are rubbed together or roll over one another to generate charges with an extremely high potential. In the Van der Graaff generator an endless plastic belt is used to carry the charge from a relatively high voltage power pack up an insulated tunnel to the metallic collection dome. The build-up of charge continues until it flashes over to the nearest earthed object. Even a small laboratory model can produce potentials of hundreds of thousands of volts.

e-mail (*comp*) See: *electronic mail.*

embossing (*mech*) A process for producing a raised pattern on sheet metal by pressing the metal between dies, one of which will have a raised impression and one which will have the corresponding hollow impression. Not to be confused with coining.

embrittlement (*matls*) Reduction of toughness in materials resulting from the process temperature and the environment in which the material is used. Examples are temper brittleness, weld decay and intercrystalline corrosive embrittlement. Ageing can also cause embrittlement in some materials, particularly plastics.

emery (*matls*) A naturally occurring abrasive found in Greece and Asia Minor. It consists of a mixture of carborundum and magnetite or haematite. Nowadays synthetic abrasives are more likely to be used.

emery paper (*mech*) Paper or cloth coated with emery powder attached to the paper or cloth by an adhesive and used for polishing or finishing metal surfaces.

e.m.f. (*elec*) See: *electromotive force.*

emissivity (*sci*) The ratio of the total energy emitted by a surface to the energy emitted by a *black body* at the same temperature, a black body having an emissivity factor of 1 and a perfect radiator an emissivity factor of 0. A polished shiny surface has an emissivity factor of approximately 0.1 whereas a surface coated with matt black paint has an emissivity factor of approximately 0.95.

emitter (*elec*) Terminal of a *bipolar transistor.*

enable pin (*elec*) Connection of an integrated circuit, used to switch the device on, i.e. to *enable* it.
endothermic (*sci*) Chemical reaction that absorbs heat from the surroundings. See also: *exothermic*.
energy (*sci*) The term 'energy' is used to describe the ability of a system or body to do work or to describe the work stored in a system or body. The units of energy are therefore the same as those of work, i.e. the joule (J). Energy is generally classified into either *potential energy* or *kinetic energy*. Potential energy is the energy stored in a system or body owing to its position or condition; this includes nuclear and chemical energy. Kinetic energy is the stored energy of a body owing to its motion, and can be defined as the work done in bringing the body to rest.
engine (*sci*) Device that converts one form of energy into mechanical work. The usual use of the term 'engine' implies a heat engine of some sort which is a system operating in a cycle which produces a net quantity of work from a heat supply; examples are the *internal combustion engine* and the *steam turbine power cycle*.
enthalpy (*sci*) Property of a fluid in thermodynamic theory, particularly relating to open systems, that is a combination of the pressure, p, volume, V, and internal energy, U. Quantity symbol is H and the units are the same as energy (J or kJ), the specific enthalpy, h, being measured in kJ/kg. Expressed mathematically, $H = U + pV$. Values of enthalpy are found in tables of properties of fluids, and as with internal energy values, the enthalpy values given are relative to a reference value.
entropy (*sci*) Measure of the availability of the energy of a system to do work, quantity symbol S; it is also considered as a measure of a system's disorder. When a system undergoes a reversible process, the change in entropy, ΔS, is equal to the heat transfer divided by the temperature at which this occurs, i.e. $\Delta S = \Delta Q/T$; hence the area under an absolute temperature versus entropy

graph of a reversible process represents the heat transfer of the process. An increase in entropy is always accompanied by a decrease in the energy available.

epicyclical gears (*mech*) A gear train consisting of a stationary internally toothed annular ring and a *sun gear* in the centre of this ring driven by the input shaft. Between the sun gear and the annular gear are several pinion gears called *planet gears* mounted on a carrier which is attached to the output shaft. When the sun gear is rotated by the input shaft, the planet gears rotate around their own axis and consequently climb around the inside of the annular ring rotating the carrier. The output shaft is driven by the carrier, at a lower speed than the input shaft.

epoxy resin (*matls*) A synthetic polymer adhesive used for high-strength joint applications and also for *potting* electrical and electronic components. Epoxy resins have very high adhesion, chemical resistance and mechanical strength.

EPROM Abb. (*comp*) Erasable programmable read-only memory, a computer 'read-only memory' which can be programmed by the user and erased by ultraviolet light.

equation (*maths*) An equation is a statement that two quantities are equal, i.e. any statement containing the sign of equality (=), e.g. $y = ax + b$. Solving an equation means finding the value of one or more unknowns from the given data. See: *linear equation*; *quadratic equation*.

equation of state (*sci*) Equation relating the pressure, p, volume, V, and temperature, T, of a quantity of gas; for example, the ideal gas law is $pV = mRT$ where R is the molar gas constant and m is the number of moles of gas.

equations of motion (*sci*) Four equations applicable to any body moving with linear motion and uniform acceleration, that relate distance travelled, s, time, t, initial velocity, u, final velocity, v, and acceleration, a. They are:

$$v = u + at$$

$$s = \frac{u + v}{2} \times t$$

$$s = ut + {}^{1}/_{2}(at^{2})$$

$$v^{2} = u^{2} + 2as$$

equilateral triangle (*maths*) See: *triangle*.

equilibrant (*mech*) A force which applied to a *concurrent force system*, places the system in a state of equilibrium. The *equilibrant* force must therefore be equal in magnitude to the *resultant* force of the system but acting in the opposite direction as shown in Fig. E.2.

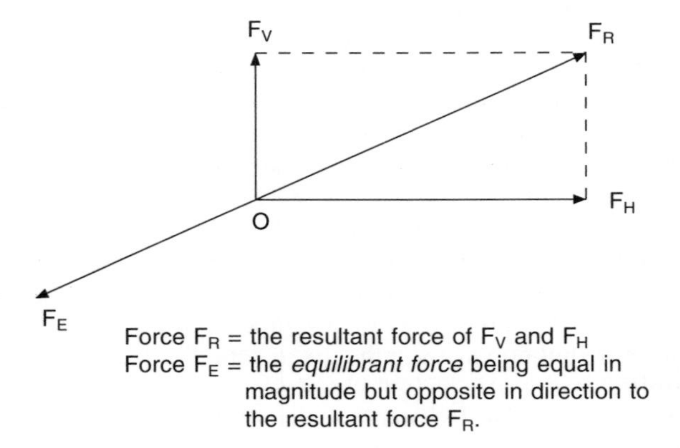

Fig. E.2 Equilibrant Force

equilibrium (*sci*) A state in which relevant quantities of a system act in such a manner as to balance each other so that there is no tendency for change. In mechanical engineering, a system is said to be in equilibrium when the effects of the forces acting on a body cancel each other out so there is no tendency for the motion of the body to change; the body can be stationary or in a state of uniform motion. A system is said to be in thermal equilibrium if there is no net heat transfer across the boundary.

erosion (*mech*) Material removal from surfaces owing to the action of liquid flow, particularly liquid containing solid abrasive particles.

error (*sci*) The difference between the actual output and the desired output of a system. In feedback control systems the error signal of a plant forms the input to the controller, so that the control system can maintain the plant output as closely as possible to the desired value, i.e. with the minimum of error. In instrumentation systems, the error is the difference between the true value of the quantity measured and the indicated value. See: *instrumentation error*.

evaporation (*sci*) Change of state of a liquid into a vapour, when molecules at the surface of a liquid have sufficient energy to break free into the atmosphere, i.e. to turn into a vapour. Sometime referred to as vaporization. Evaporation can occur at temperatures much lower than the boiling temperature owing to vapour pressure allowing some molecules to overcome atmospheric pressure. In everyday language, evaporation generally refers to the change of state occurring below the boiling temperature, such as evaporation of water at the surface of a swimming pool. See also: *latent heat of vaporization*.

excitation (*elec*) The action of *current* flowing through a coil to produce a magnetic field, especially in an *electric motors* and *generator*.

exclusive-OR gate (*elec*) (or Ex-OR gate) (*elec*) Logical, two-input operator which has an output only if the two inputs are different. If the two inputs are the same, the output is zero.

exothermic (*sci*) Chemical reaction that releases heat to the surroundings. See also: *endothermic*.

expansion 1 (*sci*) The thermal expansion of solid materials must be considered carefully in engineering design to ensure that component dimensions are satisfactory throughout the working temperature range. Also materials whose thermal expansion is restricted are subject to an increase in stress. Allowance for the thermal expansion of liquids must be made in, for example, hydraulic plant and cooling systems to prevent a rise in pressure and possible damage to equipment. Gases expand as the temperature increases for a constant pressure (see: *gas laws*) to maintain the density. A decrease in the density of a gas means that by taking a fixed volume of that gas a reduced mass is obtained. For this reason the air supply to the inlet of an internal combustion engine must be kept cool to ensure that a sufficient charge is admitted into the cylinders. Engines fitted with turbocharger systems benefit from intercoolers to reduce the density of the gas. **2** (*maths*) The writing out of a mathematical expression as a series of terms.

expectation (*stats*) The expectation of an event happening (E) is defined in general terms as the probability of the event occurring multiplied by the number of attempts made ($E = pn$).

expert system (*comp*) Search tree of 'if–then' rules in *artificial intelligence* based on expert knowledge. Comprises a knowledge base of expert knowledge and an inference engine which is the scheme of reasoning and reaching conclusions.

exponent (*maths*) The value to which a number is raised. For example, x^n indicates that x is raised to the power of n, where n is the *exponent*. The exponent function of electronic calculators or computer software packages, usually multiplies a value by ten to the power of another number; this is useful for dealing with the metric system that uses a prefix with each unit to denote multiples and submultiples of 10^3.

exponential functions (*maths*) Any function containing e^x, where e is base of natural logarithms (2.7183). The exponent arises from the natural laws of growth and decay and are of the general form $y = Ae^{kx}$ where A and k are constants. The natural laws occur frequently in engineering and science.

extensometer (*matls*) Device for measuring changes in the gauge length of a standard tensile test specimen, made from a material sample, when subjected to a tensile load. The measured change in the gauge length enables the strain in the test specimen to be determined for a given applied stress.
extrusion (*mech*) **1** Hot extrusion is a process of metal forming by forcing heated metal (usually brass or aluminium alloy) through a *die orifice* by a hydraulically powered ram, to produce an accurate cross-section in a single operation; bars, rods, tubes and complex sections, etc., are all extruded. **2** Cold or impact extrusion is a process of metal forming where a billet of soft, malleable metal is struck by a punch so as to force the metal between the punch and a die. Originally used for the production of toothpaste tubes and similar products. The force on the punch is applied either by a mechanical or by a hydraulic press. More recently, advances in technology have enabled components made from high-tensile steels to be cold, impact extruded, notably the components for constant velocity universal joints for front wheel drive vehicles.

F

face plate (*mech*) A work-holding device used on a centre lathe when the datum surface of the workpiece is perpendicular to its axis and/or when the workpiece cannot be held in a chuck or turned between centres. The workpiece is attached to the face plate by the use of clamps.
facing sand (*mech*) In foundry work, *facing sand* is a fine moulding sand that has not been previously used. It is the sand in contact with the pattern when making the mould, and provides a smooth surface to the mould cavity. To economize in its use, it is usually *backed-up* with moulding sand that has been reclaimed from previous moulds.
factor (*maths*) When two or more numbers are multiplied together the individual numbers are called factors. For example, 2 and 3 are *factors* of 6. Inversely, 6 is a *multiple* of 2 and 3. See: *HCF*; *LCM*.
factor of safety (*mech*) The ratio between the stress in a material at the elastic limit and the working stress to which the material may be subjected. The *safe working stress* is usually less than 50% of the stress at the elastic limit (factor of safety of 2), and is the maximum permissible stress for a component that should not be exceeded in service. The factor of safety is decided by considering the type and condition of the material, the anticipated wear and corrosion, the type of load, consequences of failure and the quality of the manufacturing process.
Fahrenheit scale (*sci*) An obsolescent temperature scale whose fixed points are derived from the boiling point of water (212°F) at standard atmospheric pressure and the freezing point of water (32°F), no longer used in science and engineering. To convert from Fahrenheit to Celsius, use the following formula:

$$C = \frac{5}{9}(F - 32)$$

where F is the temperature in degrees Fahrenheit and C is the temperature in degrees Celsius.

fan (*mech*) A rotary device for creating a flow of air, used for heating, cooling, air-conditioning, fume and dust extraction and air filtering. Alternatively a flow of air past a fan will cause it to rotate so that it can be used for measuring flow rate. The greater the flow rate the faster the fan rotates. In such an application the fan drives some form of transducer which converts rotational speed into an electrical signal that can be used to register the flow rate on an appropriate read-out device. Common types are centrifugal, axial flow and propeller fans.

fan-in (*elec*) The number of gates that can be connected to the input of an integrated circuit (IC) gate without affecting its performance.

fan-out (*elec*) The number of gates that can be connected to the output of an integrated circuit (IC) gate without causing its output to change from the specified logic 0 or logic 1 levels.

farad (*elec*) Unit of electrostatic *capacitance*, symbol F; defined as the capacitance of a system that requires a charge of one *coulomb* to raise its potential by one volt. The unit is very large and usual capacitance values are measured in μF, nF and pF.

Faraday's law of electromagnetic induction (*elec*) Faraday was a nineteenth century, British scientist who investigated the effect of a magnetic field on a conductor. The law can be considered in three parts. (1) An e.m.f. is induced in a conductor when the magnetic field surrounding it changes. (2) The value of the induced e.m.f. is directly proportional to the rate of change of magnetic field. (3) The polarity of the induced e.m.f. depends on the direction of the changing magnetic field.

Faraday's laws of electrolysis (*elec*) (1) The amount of chemical change occurring during electrolysis is proportional to the charge passed. (2) The amount of different substances deposited or liberated by a fixed quantity of electricity is proportional to the chemical equivalent weights of the substances.

feed (*mech*) The penetration of the cutting tool of a machine tool into the workpiece as cutting proceeds. In a drilling machine it is the axial movement of the rotating drill into the workpiece. In a lathe it is the movement of the cutting tool against the rotating workpiece either parallel to, or perpendicular to the axis of rotation of the workpiece. Most machines provide the options of automatic (power) feed or manual feed.

feedback (*elec*) See: *closed loop*.

feedforward (*elec*) Transfer of signal of a control system from one point to another further down the system.

feed pump (*mech*) A pump that supplies a steam generator or water boiler with *feed water*. The pressure generated by the pump must be sufficient to overcome the boiler pressure and is usually of the multi-stage centrifugal type, driven by a steam turbine or electric motor. The water flow rate produced must be controlled to maintain the correct water level within the boiler drum under all steam load conditions.

feed rate (*mech*) The rate at which the tool is fed into the workpiece as cutting proceeds. In a lathe or a drilling machine the feed rate is stated in millimetres per revolution, in a milling machine the table feed rate is stated in millimetres per minute.

feeler gauges (*mech*) An instrument comprising a set of thin strips of metal of precise thickness used for accurately measuring small distances between surfaces.

ferrite (*matls*) The stable α-phase of plain carbon steel. It is a weak solid solution of carbon in iron in which the carbon content ranges between 0.006% at 20°C and 0.03% (max) at 723°C. Ferrite is relatively soft, ductile and weak.

ferrous metal (*matls*) any metal alloy containing iron.

ferrule (*mech*) **1** A gland nut for making a pipe joint. **2** The metal ring round a file handle which prevents the wood from splitting.
fettling (*mech*) Process of trimming a casting after removal from the mould.
fibre optics (*elec*) See: *optical fibres*.
fibre reinforcement (*matls*) The use of high-tensile fibres to reinforce a weaker matrix material, for example the use of glass fibres to reinforce polyester resins. Stronger but more expensive fibres such as carbon fibres are used to reinforce products ranging from fishing rods to racing-car body shells where a high strength-to-weight ratio is required.
field-effect transistor (FET) (*elec*) Unlike a bipolar junction transistor (BJT) an FET is often called a unipolar device because the current flowing through it consists of only one type of charge carrier (electrons or holes but not both). Unlike a BJT, where conduction is current-controlled, an FET is voltage-controlled by an electric field with a gate that draws no current. They are suitable, therefore, for use in amplifiers where a high input impedance is required. There are basically two types of field-effect transistors: the junction gate transistor (JFET) and the metal oxide silicon transistor (MOSFET). As many MOSFETs can be constructed in a small space, they are suitable for large-scale integrate circuits such as *microprocessors*.
filament (*elec*) High-resistance wire made from a refractory metal (e.g. tungsten) that produces heat and light when a current is passed through it; used in light bulbs.
file (computer) (*elec*) Named data set which may be stored and retrieved by the operating system of a computer.
file (hand) (*mech*) Tool for manually removing metal and generally producing flat or curved surfaces according to required dimensions. A file consists of a hardened and tempered blade with teeth and a tough tang with a wooden handle. There are

many types of file with differing teeth and sectional forms (e.g. square, flat, half-round). The *cut* of the file indicates the coarseness of its teeth. A rough-cut file is used for maximum metal removal but leaves the poorest surface finish. Files of diminishing roughness are: bastard-cut, second-cut, smooth-cut, dead-smooth-cut. The use of the file (filing) is generally considered to be the most important and the most difficult manual skill to achieve in a workshop.

filler rod (*mech*) Metal rod for filling a joint when fusion welding. See: *fusion welding*.

fillet (*mech*) The radius formed where two surfaces of a component meet. Sharp intersections are generally a point of weakness, and can be considered as an *incipient crack* as they create stress concentrations leading to failure in service. By introducing a generous radius at the intersection, particularly on castings, the strength of the component is greatly increased and it is far less prone to fatigue failure in service.

fillet weld (*mech*) See: *fusion welding*.

filter 1 (*mech*) Device for separating solid particles from a fluid. Filters are widely used in engineering to remove solids from fuels, coolants, lubricating oils, hydraulic oils, air and gases prior to the fluids entering system stages that could result in damage or increased wear of equipment. Filters are used to remove harmful deposits from air for human consumption at a workplace or within air conditioning systems. **2** (*elec*) Circuit networks based on resistors, capacitors and inductors and piezoelectric crystals designed to pass or reject particular bands of frequencies, or for removing unwanted components, noise or disturbances from a signal prior to processing; this can be *analogue* or *digital*.

fins (*mech*) **1** Fins are thin unwanted projections on a casting owing to parts of the mould not fitting together properly; for example, core prints that are slack in the mould owing to excessive shrinkage of the core sand during baking. **2** Thin projections from the cylinders of air-cooled internal combustion engines and air compressors to increase the surface area and aid cooling.

first moment of area (*maths/sci*) Mathematical concept used in mechanics; it is the product of the magnitude of an area and the distance of the centre of mass of the area from the axis, useful for finding the *centroid* of shapes.

first-order system (*maths*) Common dynamic system that can be modelled by a first-order equation or transfer function. The response to a step input to a first-order system is an exponential curve, with no overshoot, and the time response is described by the time constant, τ (Greek letter tau).

fixed block (sequential) format system (*CAE*) In this obsolescent system the instructions in each block (line of information) are always written and recorded in the same fixed sequence. No letter commences each word, but the letter is implicit from the position of the word in the block or sequence. For example, a block could read:

30 1 45.0 37.350 80 1250 2 6

Since the first word is 30, the controller reads this as N30 because of its position and, since the third word is 45.0 the controller reads this as X45.0 because of its position, and so on. Thus all the instructions have to be given in each block, including those instructions that have not changed from the preceding block, to ensure each word is in its identifying position within the sequence. See: *word address system*.

flame (*mech*) The visible component of the rapid oxidation (burning) of a fuel gas in air or oxygen. Generally it can be considered that only gaseous fuels burn with a flame. A notable exception is the burning of the metal sodium in the gas chlorine to produce sodium chloride (common salt) under laboratory conditions.

flame-cutting (*mech*) A thermal process for cutting steel. A special blowtorch is used which has a coaxial nozzle. An outer ring of flame, produced by burning an oxyacetylene mixture, raises the metal to its reaction temperature. At this point a stream of oxygen, central to the heating flame, is directed onto the heated metal. Rapid oxidation takes place and the metal burns away. Once started the reaction is exothermic, the reaction temperature is largely self-maintaining and cutting can proceed by moving the cutting-torch along a prescribed path either manually or by using some form of mechanical guidance.

flame-retardants (*matls*) Substances added to materials to reduce their flammability (increase their fire resistance).

flange (*mech*) A projecting collar fitted to the end of pipes or shafts to support nuts and bolts for coupling the pipes or shafts together.

flash (*mech*) **1** A thin wafer of metal produced at the edges of a component whilst forging as a result of metal squeezing out between the faces of the dies. The flash is clipped from the forging before further processing. **2** A thin wafer of metal produced at the edges of a component whilst pressure die-casting as a result of metal squeezing out between the faces of the dies. The flash is clipped from the casting before further processing. **3** A thin wafer of plastic material produced at the edges of a component whilst moulding as a result of plastic material squeezing out between the surfaces of the moulds. The flash is removed from the moulding before further processing.

flask (*mech*) The complete moulding box comprising the cope and the drag as used for making sand moulds in foundry work. See: *cope*; *drag*.

flatness (*mech*) A measure of the approximation of a surface to a true plane. A workshop test of surface accuracy can be carried out by comparison with a *surface plate* of superior accuracy to the surface being tested. An oil-based marking 'blue' is lightly smeared on the surface plate. The workpiece is then placed on the surface plate and moved around. Marking blue will be picked up by the high spots of the workpiece surface. The flatness can be assessed from the pattern of coloured marks left on the workpiece surface; an even distribution indicates flatness.

flatter (*mech*) Hand forging tool for finishing flat surfaces, with a flat underneath surface approximately 75 mm square and a short shank with an upper striking surface for hammer blows. A handle is fitted to the shank of the flatter to keep the user's hand away from the point of impact of the hammer.

Fleming's rules (*elec*) Rules to determine the direction of the field, current and force in electrical machines. The first finger, the second finger and the thumb are held at right angles to each other as shown in Fig. F.1. The thumb indicates the direction of motion, the first finger indicates the direction of the magnetic field and the second finger indicates the direction of the current flow. The left hand is used for motors and the right hand is used for generators.

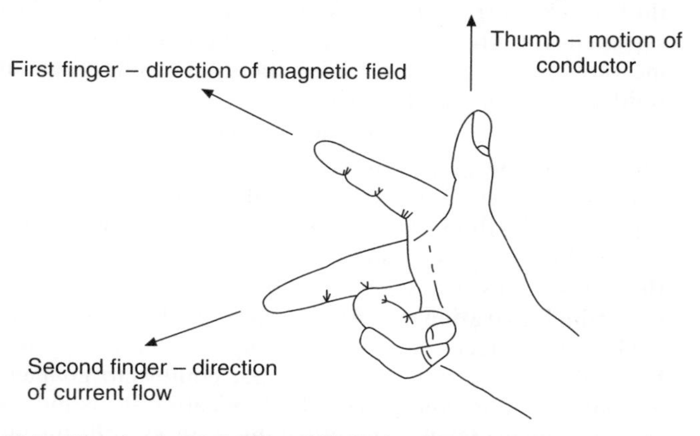

Fig. F.1 Fleming's Right Hand Rule for Generators
Note: The equivalent left hand rule is for motors.

The thumb indicates the direction of movement of the conductor relative to the flux, and applies whether the conductor moves and the field is stationary, or if the conductor is stationary and the field moves.

flip-flop (*elec*) Any circuit that can be in either of two states, reversed by a pulse. See: *bistable circuit*; *monostable circuit*; *multivibrator*.

floppy disk (*comp*) Device used for non-volatile storage of data for personal computers. The disk is very thin and coated with a magnetic material. Data is stored on the disk in concentric circular tracks. The disk is housed in a hard envelope, and data is read from or written to the disk through a slot in the envelope using an electromagnetic read/write head.

flow rate (*sci*) The flow rate of a fluid can take three different forms: mass flow rate ($\dot{m}$) measured in kg/s, volume flow rate ($\dot{V}$) measured in m^2/s and fluid velocity (c) measured in m/s.

flow work (*sci*) The work done in moving fluid through a system as it is propelled by the fluid behind it. This is also called the flow energy or the pressure energy. For one kilogram of a fluid moving through a system

$$\text{flow work} = pv$$

The rate of energy entering or leaving a system owing to flow work is $\dot{m}pv$ where $\dot{m}$ is the mass flow rate of the fluid, p is the pressure of the fluid and v is the specific volume of the fluid. See: *steady-flow energy equation*.

fluid (*sci*) Any gas, vapour or liquid.

fluid film lubrication (*mech*) A state of perfect lubrication where the bearing surfaces (shaft and bearing) are completely separated by a film of oil so that they cannot come into contact under normal operating conditions. The lubricating oil is induced and sustained by the relative motion of the surfaces or by the injection of oil under pressure by an oil feed pump. See: *boundary lubrication*.

fluidized bed combustion (*mech*) Furnace that uses a fluidized bed of pulverized coal particles suspended in an air steam. Fluidized bed furnaces can operate at lower temperatures than conventional furnaces, thus reducing the formation of nitrogen oxides whilst maintaining complete combustion; they are therefore less polluting.

flux (*mech*) A substance applied to the surface of a metal prior to soldering or brazing. **1** An *active* flux reacts with the surface of the metal to remove any oxide film as well as protecting the joint zone from atmospheric oxidation whilst the joint is heated. A typical example is an acidulated zinc chloride solution. Unfortunately, active fluxes leave a corrosive residue that must be thoroughly washed off after the joining process is complete. Therefore active fluxes are unsuitable for fine electrical and electronic work and for sealing food cans. **2** A *passive* flux has no cleaning action but merely protects the surface during the soldering or brazing process. These fluxes are suitable for electrical and electronic work and for sealing food cans. Typical examples of passive fluxes are resin, tallow and olive oil. **3** Borax-based fluxes are used for brazing. All fluxes have the additional function of 'wetting' the joint surfaces so that the molten solder or molten brazing spelter will run evenly between the joint surfaces. **4** *Electric arc welding* uses flux-coated electrodes; the flux produces a protective bubble of carbon dioxide around the joint as it burns and leaves a protective layer of slag over the weld. It also stabilizes the electric arc. **5** (*elec*) See: *electric flux*; *magnetic flux*.

flux density (*elec*) **1** Electric flux density (D) measured in coulomb/metre^2 is defined as the magnitude of the electric flux (ψ) measured in coulombs, per area (A) of the electric field measured in m^2 at right angles to the lines of force

$$D = \frac{\Psi}{A}\ (\text{C/m}^2)$$

Sometimes the symbol of charge, Q, is used instead of ψ. Thence

$$D = \frac{Q}{A}\ (\text{C/m}^2)$$

See also: *electric flux*. **2** Similarly, magnetic flux density (B), measured in tesla, T, is defined as the magnitude of the magnetic flux (Φ) measured in weber, Wb, per area (A) of the magnetic field measured in m^2 at right angles to the lines of force:

$B = \dfrac{\Phi}{A}$ See also: *magnetic flux*.

fly cutter (*mech*) Simple milling cutter in which a single-point tool is inserted into a suitable holder as shown in Fig. F.2.

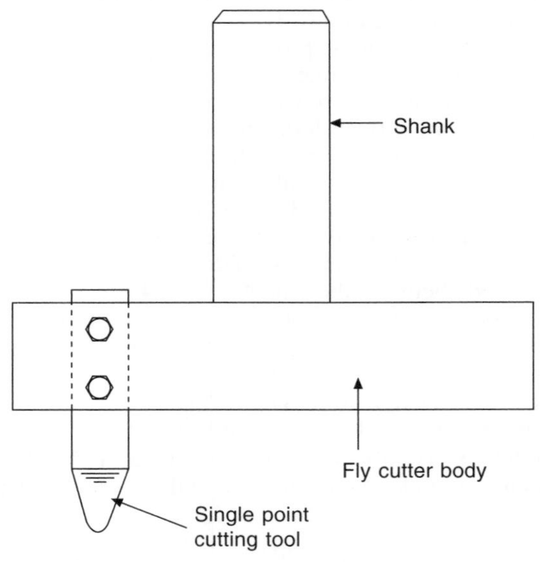

Fig. F.2 Fly Cutter

fly press (*mech*) Simple press in which the ram is raised and lowered by rotating a multi-start screw by means of a handle. Heavy cast-iron balls attached to the handle provide momentum to the screw action, thus providing a considerable downward force onto the ram and, therefore, onto the tool (punch) fixed to the press ram. A typical press is shown in Fig. F.3.

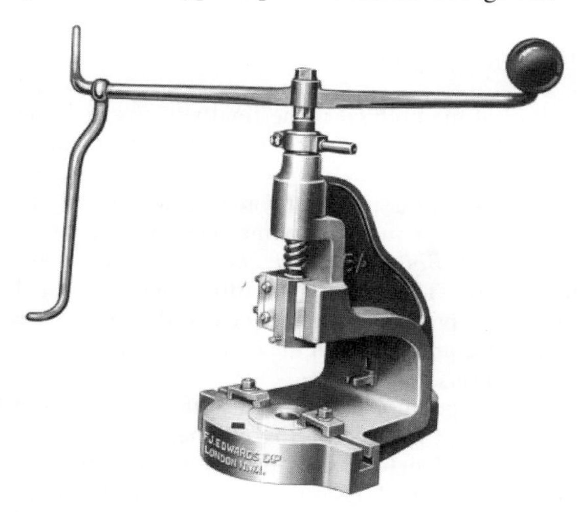

Fig. F.3 Fly Press (courtesy of 600 Group plc)

flywheel (*mech*) Component of any rotating mechanism that acts as a reservoir of kinetic energy, reducing cyclic variations of speed, and essential for the smooth running of any reciprocating engine or compressor.
follower 1 (*elec*) Terminal of *bipolar transistor*. **2** (*mech*) Component in contact with and actuated by a cam.
footprint 1 (*comp*) Commercial term for the desk space occupied by computer equipment. **2** (*mech*) A tool with easily adjustable jaws used in pipe fitting, so called because of its shape. They are used for gripping gland nuts, etc.

force (*sci*) Action that changes or tends to change the motion of a body on which it acts. The magnitude of the force (F) is proportional to the rate of change of momentum of the body on which it acts:

$$F \propto \frac{m(v - u)}{t}$$

where m is the mass of the body, v is the final velocity, u is the initial velocity and t is the time taken for the momentum change. See: *Newton's laws of motion*.
force diagram (*sci*) Vector diagram of forces drawn usually to establish the stress in structural members or mechanism components for design purposes. See: *Bow's notation*; *shear force diagram*.
forced-draught (*mech*) The use of fans to provide additional air, at above atmospheric pressure, to a furnace in order to increase the combustion rate as compared with the combustion rate under natural draught conditions.
force pump (*sci*) Any pump used to increase the pressure of a liquid in a system, e.g. the pump on a fire engine.
forge (*mech*) A complete plant for forging processes. In its simplest form it is the blacksmith's workshop comprising a hearth or furnace for heating metal, an anvil, together with the assorted hammers and hand tools of the craft of forging. On an industrial scale a power hammer, drop-stamp or a forging press together with mechanical manipulators replace the hand tools of the blacksmith.
forge welding (*mech*) Unlike fusion welding, the metal being joined (parent metal) is not melted. It is raise to a white heat just below its melting temperature, powdered flux is sprinkled into the joint, and the weld is completed as components are hammered together, the pressure causing the metals to flow. The wrought metal structure of a forge weld produces a stronger joint than the cast structure of a fusion weld. See: *spot welding*; *butt welding*; *resistance welding*.
forging (*mech*) The forming of malleable metals by plastic flow.

The metal is squeezed between tools or dies, using hammers or presses, to produce the desired shape. Usually the metal is heated to above its temperature of recrystallization to increase its malleability and reduce the chance of cracking. See: *recrystallization*; *riveting*.
format (*CAE*) See: *program format*.
formatting (*comp*) See: *disk formatting*.
form-tool (*mech*) Contoured cutting tool for producing a profile on the workpiece, e.g. cutting a radius or a screw thread on a lathe or cutting gear teeth on a milling machine.
FORTRAN (*elec*) Abbreviation of 'formula translation'; a high level scientific programming language.
fossil fuel (*sci*) Fuel derived from fossilized living organisms, i.e. coal, oil and natural gas.
foundation ring (*mech*) Rectangular iron or copper ring for securing the edges of firebox plates of a locomotive-type boiler.
foundry (*mech*) Workshop where metal objects are cast in moulds.
Fourier analysis (*maths*) This is a method of analyzing the harmonic components of complex waveforms in the terms of the Fourier series, which represents the waveform either using a wave analyzer or by mathematical treatment, where the function is represented by a trigonometric series:

$$\mathrm{f}(x) = a_0 + a_1 \cos x + a_2 \cos 2x + a_3 \cos 3x + \ldots$$
$$+ b_1 \sin x + b_2 \sin 2x + b_3 \sin 3x + \ldots$$

where a_0, a_1, a_2, a_3, …, b_1, b_2, b_3, …, are called the Fourier coefficients of $\mathrm{f}(x)$. Fourier analysis forms an important tool for the analysis of problems in engineering involving electrical, mechanical and thermal waveforms with a periodic function.
four-stroke (Otto) cycle (*sci*) One of two possible cycles of operation of the internal combustion engine, the four-stroke cycle was developed by Nikolaus August Otto in 1876. The four-stroke cycle is found in both *spark-ignition* and *compression ignition engines*. It requires four strokes for each cycle; the *induction*

stroke, the *compression stroke*, the *power stroke* and the *exhaust stroke*, completed over two revolutions of the crankshaft as shown in Fig. F.4. See: *two-stroke cycle*.

fraction (*maths*) A part of a whole, a number less than 1 but greater than 0. For example, 3 divided by 4 may be written as $(3 \div 4)$ or $^3\!/_4$, where $^3\!/_4$ is called a fraction. The number above the fraction bar (which may be horizontal as well as sloping) is called the *numerator* and the number below the fraction bar is called the *denominator*.

framed structures (*sci*) Ideal concept used in the study of frameworks in mechanics to simplify the theory. Framed structures allow the analysis of static forces by vector diagrams. A framed structure is defined as a structure built of straight bars joined at the ends by frictionless pins so each bar is in direct tension (tie) or direct compression (strut). In practice, a steel framework would usually be riveted or welded at the ends of each member. The weight of each bar is assumed to be negligible compared to the load that the framed structure carries. The framed structure is

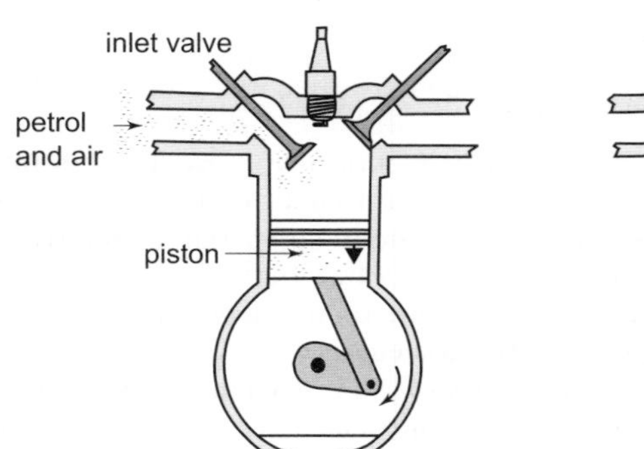

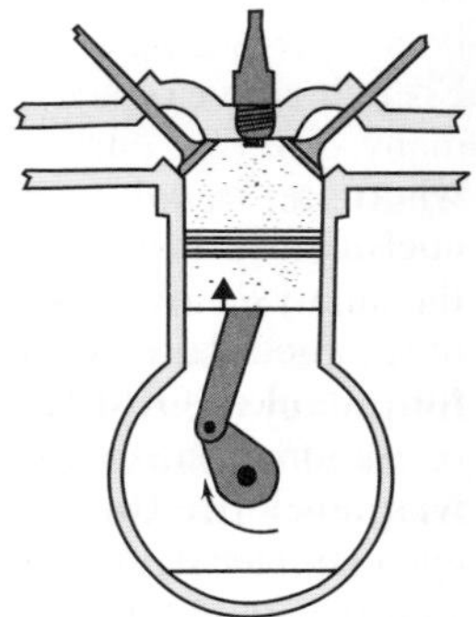

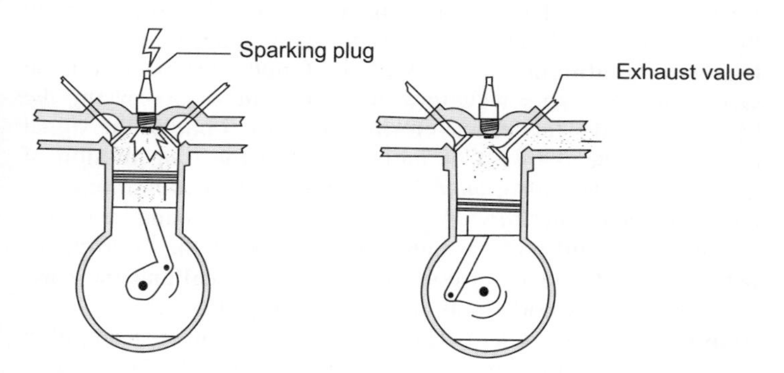

Fig. F.4 The 4-stroke Pertol Engine—Suck, Squeeze, Bang and Blow

usually supported at one end by a hinge whilst the other end is supported by rollers.

freeware (*comp*) *Software* freely available and not requiring a licence fee for use.

French curve (*mech*) Stencil drawing-office instrument that has many differing curved profiles.

frequency (*elec*) The rate of repetition of a periodic event, measured in hertz (Hz), which is the number of repetitions per second of an event. The frequency (f) of a signal is the reciprocal of the periodic time (T), where $f = 1/T$.

frequency (*stats*) The number of times a member occurs in a set is called its *frequency*. See also: *tally chart*.

frequency division multiplexing (*elec*) A technique by which one channel may be used for the transmission of multiple signals simultaneously, thus making more efficient use of the channel.

Each of the signals using the channel will have a different modulated carrier frequency. The required signal is isolated on reception by the use of tuning circuits similar to those found in a radio receiver.
frequency modulation (*elec*) Method of superimposing a periodic signal on to a carrier waveform by increasing or decreasing the frequency of the carrier waveform as the amplitude of the signal increases or decreases. Commonly used in the transmission of radio signals. See also: *modulation*; *amplitude modulation*.
frequency polygon (*stats*) A graph produced by plotting frequency against class mid-point values and connecting the coordinates with straight lines. The frequency axis is vertical and the class mid-point axis is horizontal, as shown in Fig. F.5.
frequency response (*elec*) The variation in the gain of an amplifier when plotted against variation in frequency. See: *Bode diagram*.
friction (*sci*) Force that opposes the movement of one surface relative to another with which it is in contact. *Dry friction* (F_f), also called *Coulomb friction*, applies to clean dry surfaces. As the

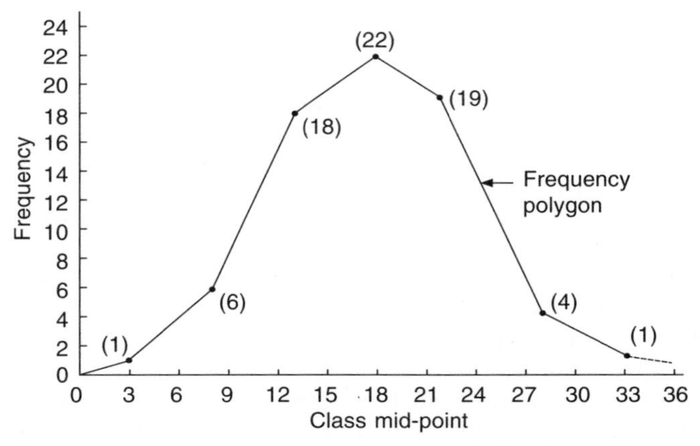

Fig. F.5 Frequency Polygon Based on Tally Chart-Fig. T.1

two surfaces move relative to one another, tiny irregularities (asperities that exist no matter how smooth the surface) interlock and local pressure welding momentarily occurs, creating a resisting force that opposes the relative motion and acts parallel to the contact surfaces. When a gradually increasing force is applied that tends to move one surface relative to another, there is an initial force that prevents motion from taking place; this is called *static friction* ($F_{f(s)}$) and will reach a maximum value just before relative motion takes place. As the interlocking asperities shear, relative motion between the two surfaces occurs. The frictional resistive force of this relative motion is called the *kinetic force of friction* ($F_{f(k)}$) which is slightly lower than the maximum (limiting) value of static friction (see: *Coulomb's laws of dry friction*). For both types of dry friction, under normal load and speed conditions, the ratio between the frictional force (F) produced and the normal contact force (R_N), is a constant called the coefficient of friction (μ). The coefficient of friction is a function of the nature of the contact materials and the condition of the surfaces and is independent of the area of contact:

$$\text{coefficient of static friction, } \mu_s = \frac{F_{f(s)}}{R_N}$$

$$\text{coefficient of kinetic friction, } \mu_k = \frac{F_{f(k)}}{R_N}$$

friction clutch (*mech*) See: *clutch*.

friction losses (*sci*) Energy losses of a machine resulting from the friction of sliding components, rolling components and the air resistance of exposed mechanisms. Since friction losses represent wasted energy they must be kept to a minimum.

fuel (*sci*) Any substance that may be burnt or chemically altered in some way so as to convert its potential chemical energy into usable heat energy. Heat engines use fuel to provide heat energy for conversion into mechanical work, either by burning the fuel internally (e.g. internal combustion engine) or externally in a furnace (e.g. steam turbine plant).

fuel cells (*elec*) These 'burn' fuel in such a way as to convert it directly into electricity without the need for an engine-driven generator. Hydrogen–oxygen cells have been used in space vehicles where the water produced as a by-product is an essential requirement.

fulcrum (*mech*) Point about which a body rotates, particularly for a lever.

fullers (*mech*) Forging tools, made in pairs, with the lower tool shaped to sit in the hardie-hole of an anvil. They are used for initial thinning and drawing-out operations resulting from their wedging action, as shown in Fig. F.6.

full-scale deflection (*elec*) An input to *instrumentation* that causes the needle of a dial to move to the maximum limit.

full-wave bridge rectifier (*elec*) A *rectifier* circuit that makes use of both half-cycles of an alternating current. It consists of four diodes connected in a bridge configuration. The direction of the current through the load resistor is constant for the whole cycle of the a.c. supply. See *half-wave rectification*; *smoothing circuit*.

fuse (*elec*) Device consisting of a length of wire with a low melting point enclosed in a glass or ceramic cylinder, incorporated in an electrical circuit, designed to fail at a specific current load for protection of equipment and the rest of the circuit. The time to failure is approximately inversely proportional to the amount by which the current exceeds the rated fuse current.

fusion (*matls*) Phase change from a solid to a liquid, e.g. the melting of metal.

fusion welding (*mech*) Any welding process where the edges of the metal components being joined (*parent metal*) are melted and run together to form a continuous joint on solidifying. Additional metal is added by the controlled melting of a suitable filler rod into the joint gap to maintain or enhance the metal thickness

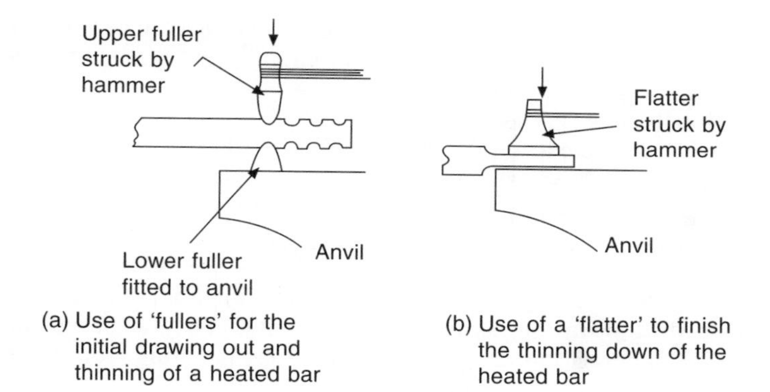

(a) Use of 'fullers' for the initial drawing out and thinning of a heated bar

(b) Use of a 'flatter' to finish the thinning down of the heated bar

Fig. F.6

across the joint to avoid weakness. Some typical welded joints are shown in Fig. F.7. See: *electric arc welding*; *gas welding*; *MIG welding*; *TIG welding*.

fuzzy logic (*comp*) A system of logic that operates on *multivalent*, rather than conventional *bivalent* logic, where everything is a matter of degree, allowing an infinite range of possibilities between two limits rather than just the two values, e.g. infinite shades of grey between black and white. This has been developed in recent years, initially by L. Zadeh in California, into a system of reasoning using fuzzy sets and fuzzy rules. A fuzzy set is a set whose members belong to it to some degree, allowing representation of properties by degree. A fuzzy rule is the relationship between fuzzy sets. In a fuzzy control system, inputs from the system under control trigger a set of fuzzy rules in a *microprocessor*, each to some degree between 0 and 1, to produce the output control signal. A fuzzy system can estimate a continuous system, and can be implemented using rules of thumb and the intuition of an experienced plant operator without mathematical models. For example, in Tokyo, Japan, a fuzzy control system has been built

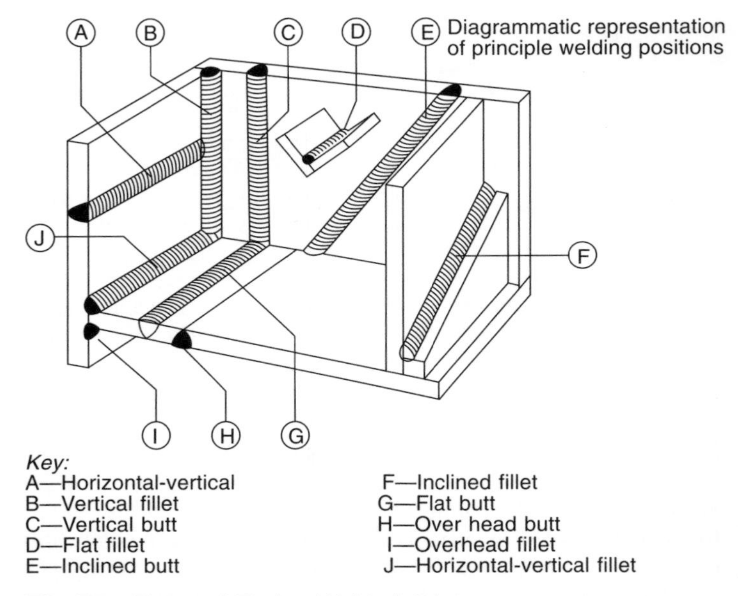

Fig. F.7 Types of Fusion Welded Joints

that can stabilize a helicopter in flight if it looses a rotor blade, a system that would not be possible to estimate using conventional mathematical models. Fuzzy systems can be adaptive using a learning system to produce or tune fuzzy rules from plant behaviour. These operate by sampling plant variables and clustering data into rules, commonly done using *neural networks*. Large-scale research programs are underway in Japan where fuzzy logic is made use of in many industrial products, including vehicles in such systems as anti-lock brakes, automatic transmission and fuel injection systems. In the West fuzzy logic is still viewed with much suspicion particularly in academia where often conventional mathematical techniques are still preferred.

G

galvanic corrosion (*matls*) See: *electrolytic corrosion.*
galvanizing (*matls*) **1** Electrolytic galvanizing is the electroplating of carbon steel sheet and finished products with a film of zinc to prevent corrosion. **2** Hot-dip galvanizing is the immersion of finished carbon steel products in molten zinc to prevent corrosion. It provides a thicker and more robust coating than electrolytic galvanizing and has the added advantage of sealing any joints.
galvanometer (*elec*) A sensitive instrument for detecting the strength and direction of a direct current flow. The current flows through a coil of fine wire suspended between the poles of a permanent magnet. Interaction between the magnetic flux fields of the magnet and the coil when the current flows tends to rotate the coil. The coil carries a mirror and any deflection of the coil moves a light beam across a scale, as shown in Fig. G.1. The light beam magnifies the movement of the mirror and increases the sensitivity of the instrument.
gantry (*mech*) Structure for carrying an overhead, travelling crane structure, usually in an industrial building or an engine room.
gap bed (*mech*) A type of centre lathe whose bed has a gap in front of the headstock to allow face plates larger than standard to be fitted for the turning of large-diameter but narrow workpieces.
gas (*sci*) State of matter in which the molecules of a substance have received sufficient heat energy to move freely, enabling them to disperse indefinitely so that they may fill any vessel in which they are contained.
gas laws (*sci*) **1** *Charles's law* states that the volume of a given mass of a gas at constant pressure is directly proportional to its

thermodynamic (absolute) temperature. Therefore all gases have the same coefficient of expansion at the same pressure. **2** *Boyle's law* states that the volume of a given mass of a gas is inversely proportional at constant temperature. Neither of these laws apply to vapours or gases containing any vapour. They only apply to dry gases. Further, there are deviations at very low and very high pressures. See: *perfect gas.*
gasket (*mech*) Shapes cut out of soft, sheet material that are used to achieve a sealed joint between two mating surfaces. Although the mating surfaces of components are machined as flat and

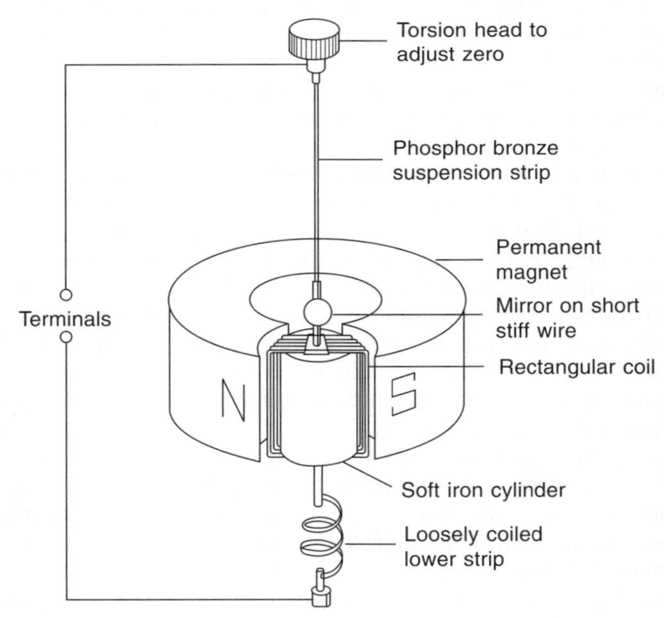

Fig. G.1(a) Main features of a suspended-coil galvanometer

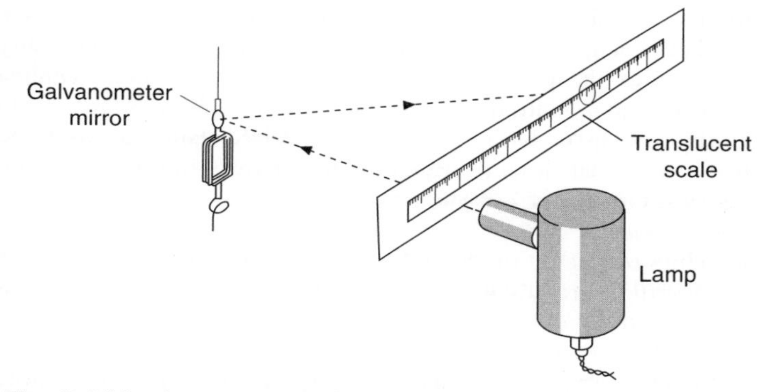

Fig. G.1(b) Arrangement of galvanometer, lamp and scale to increase the sensitivity of the galvanometer

accurately as possible, there are always irregularities in the form of roughness and waviness that prevent the two surfaces from achieving fluid-tight joints. The function of the gasket is to fill in the gaps caused by the irregularities so that the seal becomes liquid- or gas-tight even under pressure. The shape of the gasket matches the shape of the mating surfaces with holes cut for the bolts that force the two surfaces together, and provision for any fluid ducts. The gasket material and construction must be such as to resist the operating pressure and temperature and any chemical attack of the fluids.

gas welding (*mech*) Fusion welding process that uses the heat produced from the burning of a mixture of gases, usually oxygen and acetylene. Oxygen and acetylene are stored separately in cylinders, the flow of each individual gas to the torch being controlled by regulator valves. The oxyacetylene flame burns at a very high temperature (3500°C), melting the metals to be joined. A metal filler rod of similar composition to the parent metal is used to supply additional weld material and is melted into the weld pool of molten metal created by the flame. No flux is normally required as the products of combustion form a shield against atmospheric oxidation.

gauge pressure: (*sci*) Pressure measured using atmospheric pressure as the datum; most pressure gauges indicate pressure above atmospheric pressure, i.e. *gauge pressure*. See: *absolute pressure*.

gearbox (*mech*) An assembly in which a power input shaft is connected to an output shaft by means of a series of gearwheels which alter the output speed and torque to suit the plant requirements. The gears are often arranged so that the gear ratio can be changed as required by an external control lever. The whole assembly is usually contained within a sealed cast-metal box that also contains the lubricating oil. An example is the all-geared headstock of a lathe.

gear pump (*mech*) A positive displacement force pump consisting of two small-diameter coarse-pitch gearwheels in mesh with each other and enclosed in a close-fitting casing. The liquid is carried round from the suction intake to the delivery port in the tooth spaces. Such pumps are normally used for lubrication purposes where they deliver the lubricant under pressure to the bearing.

gearwheel (*mech*) A toothed wheel that rotates on a shaft and meshes with another gearwheel on another shaft to allow one shaft to drive the other at a fixed rate. The relationship between the angular displacement of the driver shaft to the driven shaft is as follows:

rev/min of the driven gear

$$= \text{rev/min of the driver gear} \times \frac{\text{number of teeth of driver gear}}{\text{number of teeth of driven gear}}$$

The simplest form of gearwheel is the straight-tooth gear where the teeth are parallel to the centre line of the gear. The helical-tooth gear consists of teeth that form a helix across the gear which are smoother and quieter than the straight-toothed type. The disadvantage of the helix type is that the gears tend to be pushed sideways on load and require special thrust bearings or a double helical teeth pattern where one helix balances the thrust of the other. See: *bevel gear*.

generation (*mech*) The machining of a profiled or contoured surface using standard cutters. This is achieved by the simultaneous and relative movement of the machine's slides, and avoids the need for expensive form cutters. Generation of a surface usually allows for more rapid metal removal without chatter occurring than is possible when using form cutters. See: *forming*.

generator (*elec*) A machine that converts mechanical energy into electrical energy through a process of electromagnetic induction. If the generator produces direct current it is referred to as a *dynamo*. If the generator produces an alternating current it is referred to as an *alternator*. Generators that supply the national grid are usually driven by a steam turbine, although gas turbines, water turbines and wind turbines are all used to a lesser extent. See also: *alternators*; *dynamos*.

geometric series (*maths*) A series of numbers with a fixed difference between terms, e.g. 1, 2, 4, 8, 16, 32, ... is geometric where each term is determined by *multiplying* the previous terms by the *common ratio* (r), which in this example is 2. If the first term is a and the common ratio is r, then the nth term is ar^{n-1}. See also: *arithmetic series*.

geometry (*maths*) That branch of mathematics in which the properties of points, lines, surfaces and solids are investigated. See: *angles*; *triangles*; *circles*.

gigabyte (*comp*) 10^9 bytes. The storage space provided by hard disks of personal computers is usually considered in gigabytes.

gradient (*maths*) The slope of a line or inclined plane to the

horizontal. The gradient of a straight line on a graph can be described by the expression $\delta y/\delta x$. See: *linear equation.*
granodizing (*matls*) See. *conversion coatings.*
graph (*maths*) A drawing representing the relationship between two variables. There are various types of graphical representation: straight-line, curves, graphs relating mathematical expressions, pie charts, histograms, bar charts and ideograms relating statistical data.
grey cast iron: (*matls*) An iron–carbon alloy in which the amount of carbon present exceeds the solubility of carbon in iron. In the solidified metal, the excess carbon appears as flakes of graphite between the grains of the metal, so that when the metal is fractured, the fractured surface is grey in colour. Although reducing the strength of the metal in tension, the presence of the free graphite gives components made from grey cast iron self-lubricating properties, which renders the material useful for machine slideways and other bearing surfaces.
grinding machine (*mech*) Any machine tool in which a rotating wheel of *abrasive* material is used to remove metal from a workpiece. Different types of grinding machine are used for various applications. For example, the *surface grinding machine* is used for producing flat surfaces to prescribed dimensions, *the cylindrical grinding machine* is used for producing internal and external cylindrical surfaces of high accuracy, *the off-hand grinding machine* is used for refurbishing single-point cutting tools as used on centre lathes.
grouped data (*stats*) A set of values obtained by forming a frequency distribution is called *grouped data*. The terminology associated with grouped data is explained in Fig. G.2.
gudgeon pin (*mech*) Component of an internal combustion engine that connects the piston to the connecting rod at the small end, allowing thrust to be transferred from and to the piston and allowing relative angular movement of the connecting rod with

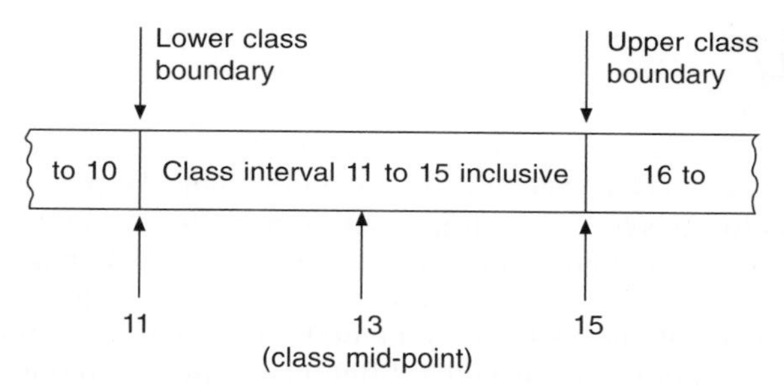

Fig. G.2 Grouped Data

crankshaft rotation. Any pin that is used in a similar application elsewhere in a mechanism.

gunmetal (*matls*) All copper/tin bronze alloys suffer from oxidation of the tin during casting. This renders the alloy brittle. To prevent or reduce the onset of oxidation a deoxidizer has to be added. This may be phosphorus or zinc. In gunmetal, zinc is used as a deoxidizer, and this also improves the flow properties, so that sound, pressure-tight castings can be produced. A typical alloy contains 88% copper, 10% tin and 2% zinc. Originally used for naval armaments – from whence it gets its name – it is now used for plan bearings or where strong, corrosion resistant, pressure-tight castings are required.

H

hacksaw (*mech*) A cutting tool comprising a metal cutting sawblade supported within and tensioned by a substantial frame to which it is attached by two pegs that fit through holes in the blade ends. Blades with different tooth formations allow for their use with various metals under different cutting conditions. Hand hacksaws are used for the removal of unwanted metal prior to filing or other workshop operations. Power-driven hacksaws are used for cutting blanks from large sections of metal prior to machining.
half-wave rectification (*elec*) Simplest form of rectifier circuit for converting an alternating current into a direct current. It consists of a single diode, that only allows current flow when the diode is forward biased and blocks the negative half of the current cycle. The direct current produced is heavily pulsed and the conversion process has a low efficiency.
Hall effect (*elec*) This effect occurs when the field of a magnet and the field produced by a flow of current interact but neither the magnet nor the conductor moves. There is still a force on the electrons within the conductor and this shifts the electrons to one side as they flow through the conductor. This results in a small potential difference being developed *across* the direction of current flow, as shown in Fig. H.1. This phenomenon is known as the *Hall effect*. Semiconducting materials such as germanium show a large Hall effect and can be used to measure the strength of magnetic fields.
hard disk (*comp*) Device used in computer systems for data storage, consisting of a disk made of metal alloy coated on both sides with a magnetic material, and stored permanently in the

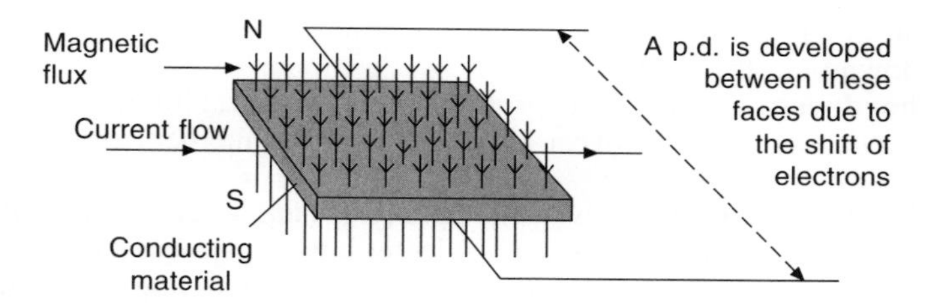

Fig. H.1 The Hall Effect

computer in a sealed dust-free environment. A hard disk is not flexible like a *floppy disk*, thus allowing for a greater amount of data storage per area of disk space, and the disk can be spun at higher speeds, allowing data to be read at a higher rate.

hardening (*matls*) Process of increasing the hardness of a material. In steel this occurs through one of the following ways. (1) By heating a high-carbon steel to above its lower critical temperature and cooling it quickly (quenching), the degree of hardness being controlled by the amount of carbon present and the rate of cooling. (2) In work hardening, cold working of the metal distorts the crystal structure, increasing the hardness and brittleness. (3) In the case of maraging steel alloys, these can be precipitation age hardened by heating to 480°C for at least 3 hours depending on the alloy. No quenching is required as age hardening results in the precipitation of hard intermetallic compounds such as TiNi in a martensitic structure from which the process and these steels get their name. See: *maraging steels*. Generally, non-ferrous metals and alloys can only be work-hardened, but a notable exception is duralumin (aluminium alloy containing copper) which is softened by solution treatment and subsequently hardens by natural ageing or more rapidly by a precipitation treatment.

hardie (*mech*) A chisel that can be located in the *hardie-hole* of an anvil with the cutting edge uppermost. The work is struck down onto the hardie.

hardness (*matls*) The property of a material to withstand scratching, abrasion, wear and indentation by another hard body. See: *hardness testing*.

hardness testing (*matls*) The testing of hardness of a material, usually by pressing a standard indenter into its surfaces by means of a standard load and measuring the extent of the resulting indentation. An exception is the Shore Scleroscope in which the rebound of a small, hard-tipped hammer is used as a measure of hardness. In all hardness tests the results are comparative. There is no absolute measure of hardness. See: *Brinell hardness test*; *Rockwell hardness test*; *Shore Scleroscope*; *Vickers hardness test*.

harmonics (*elec*) A harmonic is a sine wave which has a frequency that is a multiple of the sine wave of the fundamental frequency. Analysis of any complex waveform shows that it may be formed by the addition of sine waves of different frequencies, amplitudes and phase relationships. The fundamental is the sine wave that has the same frequency as the complex waveform and the harmonics are the remaining sine waves which have frequencies that are multiples of the fundamental. The sine wave that has a frequency of twice that of the fundamental is called the second harmonic, the sine wave that has a frequency of three times that of the fundamental is called the third harmonic, etc. The higher the order, usually the smaller the amplitude. See: *Fourier analysis*.

HCF Abb. (*maths*) Highest common factor. This is the highest number that divides into two or more other numbers exactly. Consider 12 and 30, the numbers 2, 3, 6 will all divide exactly into the given numbers and are thus all common factors, but since 6 is the largest, it alone is the HCF.

head of liquid: (*sci*) A method of expressing the pressure of a liquid. The pressure p acting on a body immersed in a liquid at a depth of h metres can be calculated from the expression: $p = \rho g h$, where ρ is the fluid density and *g* is the acceleration due to gravity. Thus the pressure head is directly proportional to the depth of the liquid and independent of the surface area or volume considered. 'Head of liquid' is often used in hydraulic terminology to describe the pressure duty of a pump. As a rough guide, one bar of pressure is approximately equal to 10 metres of water. See also: *barometer*.

headstock (*mech*) A sub-assembly at the left end of a centre lathe consisting of a cast-iron box-shaped casting supporting the *spindle* and drive mechanism that rotates the spindle. See: *centre lathe*.

hearth (*mech*) Equipment in a forge for raising the metal to its forging temperature, nowadays often gas-fired, but traditionally coke-fired with air applied under pressure from a blower or a fan, the air inlet being water-cooled (bosch). A high, localized heat is produced over which a skilled blacksmith has close control.

heat (*sci*) Energy in the process of transfer between a system and its surroundings owing to a temperature difference. Quantity symbol Q and units the same as for any other form of energy, i.e. the *joule* (J). The convention for the sign of the transfer is usually that heat entering a system from its surroundings is *positive* and heat leaving a system into its surroundings is *negative*. Heat is not a property of a system and is analogous to *work* in that it only takes place across the boundary of a system while a change occurs; to say that a system contains heat is incorrect. The term 'heat transfer' is often used to indicate an energy transfer process occurring due to a temperature gradient. Previously, units used for heat were the *kilocalorie* (*kcal*), *the British thermal unit (Btu)* and the *Celsius heat unit* (*Chu*).

heat engine (*mech*) See: *engine*.

heat-exchanger (*sci*) Device that allows the transfer of energy between a hotter fluid stream and a cooler fluid stream, where the two fluids are separated by a heat transfer surface. The fluids may flow in parallel or in counterflow. The heat transfer rate, $\dot{Q}$, between the two fluids can be calculated from the relationship $\dot{Q} = U \cdot A \cdot \Delta T_{lmtd}$, where U is the overall heat transfer coefficient, A is the area of the heat transfer surface and ΔT_{lmtd} is the logarithmic mean temperature difference which can be calculated as follows:

$$\Delta T_{lmtd} = \frac{\Delta T_0 - \Delta T_i}{\ln\left(\frac{\Delta T_0}{\Delta T_i}\right)}$$

where ΔT_i is the temperature difference between the two fluids at the hotter fluid inlet end of the heat exchanger and ΔT_0 is the temperature difference between the two fluids at the hotter fluid exit end of the heat exchanger. The formula is applicable for both parallel and counterflow arrangements although the value is higher for counterflow than parallel, so a heat-exchanger run in counterflow produces a higher heat transfer rate than the same device run in parallel. A common form of exchanger is the tube type consisting of a bundle of tubes, through which one fluid flows, in a casing, the other fluid flowing between the tubes and the casing wall.

heat-treatment (*matls*) Process of altering the properties of a metal by controlled heating and cooling. Different metals and alloys respond differently to such processes and the appropriate treatment must be used for each metal or alloy and the properties it is required to impart. For examples of heat treatment processes see: *hardening*; *tempering*; *annealing*; *normalizing*; *solution treatment*; *precipitation treatment*.

henry (*elec*) Unit of self-inductance (H). A circuit has a self-inductance of 1 H if an emf of 1 V is induced in it when the circuit current changes at a rate of 1 A/s.

hermaphrodite caliper (*mech*) Instrument with one caliper leg and one scriber leg as shown in Fig. H.2. It is used to scribe a line parallel to an edge or an arc parallel to a rim. The caliper leg is moved in contact with the edge and a line is marked with the scriber point. Also called an odd-leg caliper or a jenny caliper.

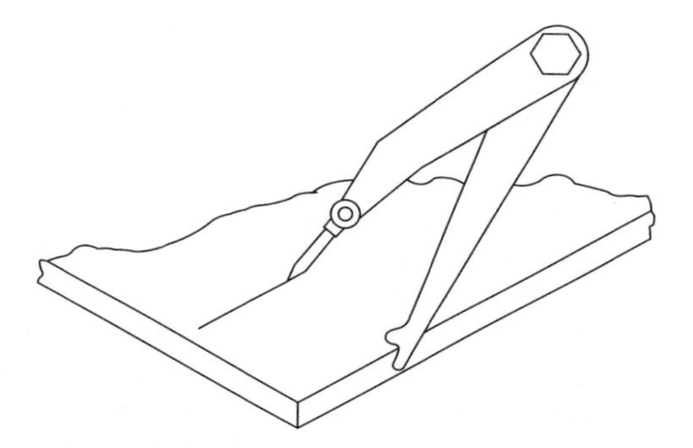

Fig. H.2 Hermaphrodite Calipers

hertz (*elec*) Unit of frequency (Hz); 1 cycle/second = 1 Hz.
heuristic (*maths*) Analysis by trial, error and logic rather than by reliance on formulae.
hexadecimal (*maths*) A base-16 numbering system, in which each digit varies between 0 to 15_{10}, and moving to the left, each digit is worth sixteen times as much as the digit immediately to its right. Digits 0–9 are the same as in the denary system (base 10) and values 10–15 are represented by the letters A, B, C, D, E and F. As with the octal numbering system, hexadecimal is useful for representing digital numbers, as conversion to or from binary is easy; a single hexadecimal digit has the same range as a four-bit binary number. For example, to convert $0101\ 1010\ 0001\ 1111_2$ to hexadecimal form: 0101→5, 1010→A, 0001→1, 1111→F. Thus $0101\ 1010\ 0001\ 1111_2$ is equivalent to 5A1F.
high-level language (*comp*) Language for writing computer programs based on English-like expressions rather than machine code or assembly language.
high-speed steel (*matls*) A group of alloy tool steels designed to

operate continuously at high cutting speeds and elevated temperatures that would soften plain carbon steels and low tungsten alloy steels. There are three groups of alloys: (1) Tungsten–chromium alloys, known as *high-speed steels*; (2) tungsten–chromium–cobalt alloys known as *super-high-speed steels*; (3) molybdenum alloys (now virtually obsolete) known as economy-high-speed steels since the cheaper and more readily available alloying element molybdenum is substituted for the more expensive tungsten and cobalt. For these reasons it was introduced during WW2.
High-speed steels are hardened by quenching them from a very much higher temperature than that required for plain carbon steels. They are then reheated and quenched in a secondary hardening process that increases both their hardness and their toughness, unlike the tempering of plain carbon steel that reduces the hardness in order to impart toughness.
histogram (*stats*) A graphical representation of class frequencies (plotted vertically) against class interval (plotted horizontally) so that the value of frequency is proportional to the *area* of the corresponding rectangle, unlike a vertical bar chart where the numerical value is proportional to the *height* of the bar or rectangle.
hogging (*mech*) Term used to describe the way that a beam is bent when it is forced up in the middle and down at the ends, as opposed to sagging. Hogging is usually considered as a negative deflection in bending moment calculations. See: *sagging*.
Hooke's law (*sci*) Robert Hooke established experimentally that, for a solid material, the strain produced is proportional to the applied stress provided that the limit of proportionality is not exceeded. This relationship is referred to as Hooke's law. See also: *elastic limit*; *tensile test*; *Young's modulus of elasticity*.
Hooke's universal joint (*mech*) Device used for connecting two drive shafts that are not parallel, consisting of two forks at right angles to each other connected to the two shafts being coupled. A crosspiece hinges the two forks together, allowing the driver

shaft to transmit torque to the driven shaft positioned at an angle to the driver shaft. Two Hooke's universal joints arranged in series may be used to transmit torque between two shafts that are parallel but not lying in the same plane.
horsepower: (*sci*) A unit of power equivalent to doing work (or using energy) at the rate of 33 000 ft lbf per minute. Largely superseded by the watt (W) for both mechanical and electrical engineering. See Appendix 7.
humidity (*sci*) Measure of the water vapour content in air. Relative humidity is the measure commonly used for industrial processes and air conditioning, and is the ratio of the moisture in the air, to the moisture it would contain if it were saturated at the same temperature, usually expressed as a percentage. Specific humidity is the ratio of the mass of water vapour to the mass of dry air in a given volume.
hydraulic accumulator (*sci*) Component of hydraulic systems that stores the energy from a fluid under pressure. Used for a variety of reasons, e.g. to provide a high flow rate over a short time period; to act as a damper to smooth out pressure pulses from pumps and pressure surges; emergency storage of energy, e.g. aircraft undercarriage; and to compensate for thermal expansion.
hydraulic head (*sci*) See: *head of liquid.*
hydraulics (*sci*) Originally referring to systems and devices operated by pressurized water, hence the name, but now refers to the study of fluids at rest (*hydrostatics*) or in motion (*hydrodynamics*). The term is generally used in engineering to refer to the fluid power transmission and control systems used throughout industry. Such systems contain pressurized oils or other liquids, and take advantage of the pressure-transmission properties of fluids. Despite fluid power systems requiring expensive equipment, being vulnerable to dirt and presenting a fire risk in the event of a leak, there are many advantages of using fluid power systems over electrical systems. These include the production of very high forces

and torques by hydraulic motors and rams, with high power-to-weight ratio hydraulic systems that are not limited by magnetic saturation as are electrical motors. Hydraulic systems lend themselves to precision speed control, high mechanical stiffness and good inching capabilities. Further, hydraulic oils carry away heat allowing a reduced component size compared to an equivalent electrical unit. Spark-free operation makes hydraulic equipment ideally suitable in such hazardous environments as coal mines and oil refineries.

hydrodynamics (*sci*) See: *hydraulics.*

hydrometer (*sci*) Instrument for measuring the density or relative density of a liquid, usually in the form of a glass tube with a long bulb at the bottom containing weights. The hydrometer floats vertically, at a depth depending upon the density of the liquid, and the reading is taken from a scale on the stem of the float at the liquid level.

hydrostatics (*sci*) See: *hydraulics.*

hyperbolic functions (*maths*) These are functions associated with the geometry of the hyperbola (see: *conic sections*):

$$\text{the hyperbolic sine of } x = \sinh x = {}^1/_2(e^x - e^{-x})$$

$$\text{the hyperbolic cosine of } x = \cosh x = {}^1/_2(e^x + e^{-x})$$

$$\text{the hyperbolic tanh of } x = (\sinh x) / (\cosh x)$$

$$= (e^x - e^{-x})/(e^x + e^{-x})$$

hysteresis (*elec*) Relationship between two quantities that is different depending on whether one is increasing or decreasing relative to the other (from a Greek word meaning to lag behind); this usually represents an energy loss to the system. A common example of this is a plot of the magnetic flux density, B, of a magnetic material against the applied field strength, H, as shown in Fig. H.3.

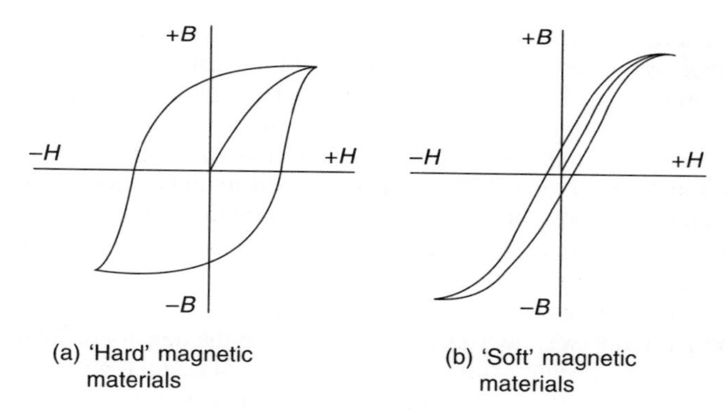

Fig. H.3 Hysteresis Loops for Magnetic Materials

I

ideographs (*maths*) Use pictorial symbols to represent quantities, often used to present statistical data to the general public as shown in Fig. I.1. See: *pictograms*.

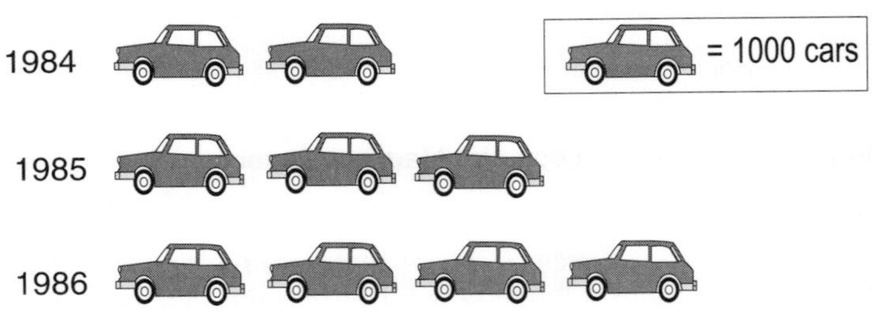

Fig. I.1 Ideograph (Pictogram)

idler (*mech*) A gearwheel in a gear train that is inserted between and meshes with the driver and driven gearwheels in order that they may rotate in the same direction. The number of teeth of the idler does not affect the *velocity ratio* of the overall train.
impedance (*elec*) The total opposition to the flow of an alternating current in a circuit. The symbol is Z and the unit is the ohm (Ω). The total opposition to flow of an alternating current in a circuit consists of the ohmic resistance, R, and the reactance, X. The complex impedance is $Z = R + \mathbf{j}X$ where $\mathbf{j} = \sqrt{-1}$. The resistance

represents loss of power and the ratio of X to R represents the phase difference between the voltage and the current. The impedance Z is also called the *apparent resistance* of the circuit, where $Z = V/I$.

impeller (*mech*) The rotor of a centrifugal pump or an air compressor.

improper fraction (*maths*) A fraction in which the numerator is greater in value than the denominator. See: *fraction*.

impulse (*sci*) Mechanics term for the product of a force and the time for which the force acts. It is equal to the change in momentum of a body, i.e. $F \cdot t = m_{2 \cdot v} - m_1 \cdot u$, where u is the initial velocity and v is the final velocity. If the mass of the body remains constant then the formula becomes $F \cdot t = m(v - u)$.

indefinite integral (*maths*) When a function containing a constant term is differentiated, the constant term is lost. Therefore when the process is reversed by integration the constant is unknown and an arbitrary constant (C) is included in the answer. Since there is no definite solution without further information, this is known as an *indefinite integral*. See: *definite integral*.

independent event (*stats*) An independent event is one in which the probability of that event occurring does not affect the probability of another event happening.

index (*pl.* indices) (*maths*) Small numbers written to the right and above a number to indicate the number of times that number has to be multiplied by itself. For example:

$$2^3 = 2 \times 2 \times 2 = 8$$

See: *powers*.

indicated power (*sci*) The power measurement in watts (W) of any reciprocating engine, calculated from the expression:

$$\text{indicated power (i.p.)} = P_m LAN$$

where P_m is the indicated mean effective pressure (N/m^2), L is the length of stroke (m), A is the piston area (m^2), and N is the rotational speed (rev/s).

indices, laws of (*maths*) When simplifying calculations involving indices there are certain rules (laws of indices) that must be applied.

- When multiplying numbers having the same base the indices are added, e.g. $2^3 \times 2^4 \times 2^5 = 2^{12}$.
- When dividing numbers having the same base the indices are subtracted, e.g. $2^5 \div 2^3 = 2^2$.
- When a number raised to a power is raised to a further power, the indices are multiplied, e.g. $(5^5)^2 = 5^{10}$.
- When a number is raised to the power of 1 it remains unaltered, e.g. $5^1 = 5$.
- When a number is raised to the power of 0 it becomes 1 (unity), e.g. $5^0 = 1$.
- When a number is raised to a negative power it becomes the reciprocal of that number to a positive power, e.g. $4^{-2} = 1/4^2$.
- When a number is raised to a power that is a fraction, the numerator is the power and the denominator is the root of that number.

induced e.m.f. (*elec*) See: *electromagnetic induction.*

inductance (*elec*) The property of a conductor which, when carrying an electric current, causes it to create a magnetic field and store magnetic energy. The magnitude of this effect is given the quantity symbol L, and the unit is the henry (H). See: *self-inductance*; *mutual inductance.*

induction motor (*elec*) An a.c. electric motor in which an electric current is passed through the stator windings from the mains supply in order to create a magnetic flux field. This *induces* an electric current in the rotor conductors which produce a second flux field. The interaction between the magnetic fields causes the rotor to rotate.

inductor (*elec*) Passive electrical component, consisting of a coil (solenoid) having a specific value of self-inductance. The energy, W, stored in a conductor can be calculated from: W (joules) = $\frac{1}{2} LI^2$, where I is the current and L is the inductance.

industrial robot (*CAE*) Industrial robots are computer-controlled, reprogrammable mechanical manipulators with several degrees of freedom, capable of being programmed to carry out a variety of industrial operations such as loading and unloading machines, welding and spray painting. See: *Cartesian coordinate robot*; *cylindrical coordinate robot*; *spherical (polar) coordinate robot*; *revolute (angular) coordinate robot*; *robot end effector*; *robot programming methods*.

inert gas welding (*mech*) Any fusion welding process in which the weld zone is protected by an inert gas to prevent atmospheric oxidation of molten metal. Carbon dioxide and argon are the most commonly used inert gases but helium, which is more expensive, is also sometimes used. See: *TIG welding*; *MIG welding*.

inertia (*sci*) Property of matter that causes it to resist a change of motion. Thus a body remains at rest unless acted upon by an external force and a body moving in a straight line at a constant velocity will continue to do so unless acted upon by an external force.

ingot iron (*matls*) Refined pig-iron cast into ingots of a size that are easy to handle and load into cupola furnaces at the foundry where it is the basic material for making iron castings. The high-purity cast iron is mixed with selected steel scrap to give the required composition and properties.

instrumentation Equipment for measuring quantities, which operates by monitoring a variable proportional to the quantity to be measured, converting the signal into a more suitable form, the reading then being recorded or displayed or used in an automatic control loop. The output may be analogue or digital.

insulator A substance that is a poor conductor of heat or electricity. With the exception of carbon, all non-metals are insulators owing to the formation of their molecular structure.

integral calculus (*maths*) Integration is the inverse of differentiation and is used to find the value of a variable whose differential coefficient is known. This mathematical process, known as integration, is used in the solution of such problems as finding the length of a curve, the area under a curve, and the volume enclosed by surfaces of revolution. In particular, the general solution of $\int ax^n\,dx$, where a and n are constants, is given by

$$\int ax^n\,dx = [(ax^{n+1})/(n + 1)] + C$$

where C is the constant of integration.
integrated circuit (*elec*) An electronic circuit built on a single chip of semiconductor material. Such chips are used extensively in computer and communication systems. They can consist of very many transistors or other components making up a complex circuit on a very small chip. Integrated circuits have largely replaced discrete component circuits owing to their low cost, high reliability and small size.
intercooler (*mech*) A liquid or air-cooled heat-exchanger used in multi-stage air-compressors to cool the compressed gas and increase its density between the compression stages, in order to reduce the work done on the gas in an attempt to get closer to an ideal isothermal compression. In *supercharged* reciprocating internal combustion engines, intercoolers may be used to increase the density of the compressed air or fuel–air mixture, allowing a greater mass to be inducted into the engine cylinders.
interference fit (*mech*) The fit achieved between mating components such as a shaft and a bush when the diameter shaft under minimum metal conditions is slightly larger than the bore of the bush also under minimum metal conditions. That is, the dimensional tolerances on the components is such that the smallest diameter shaft is still slightly larger than the largest diameter bore (hole).

internal energy (*sci*) Property of thermodynamic systems, quantity symbol U and energy unit the joule (J), considered to be the sum of the potential energies and kinetic energies of the atoms and molecules. For a closed system $\Delta U = Q - W$, where Q is the heat transfer and W is the work transfer, i.e. the heat supplied to a closed system is equal to the increase of internal energy plus the work done by the system; this is known as the non-flow energy equation.

investment casting (*mech*) A casting process in which a ceramic mould is built up around a wax pattern. When the mould is 'fired' to harden the ceramic material, the wax melts out of the mould leaving a cavity the shape of the required component. The molten metal is then poured into the mould as in any other gravity-casting process. The finish and accuracy of the castings is of a very high order and little finishing is required. Unlike die-casting, investment casting is suitable for making components from metals with high melting temperatures.

isentropic process (*sci*) Any process occurring in which there is no change in *entropy*.

ISO metric screw thread (*mech*) The international system for

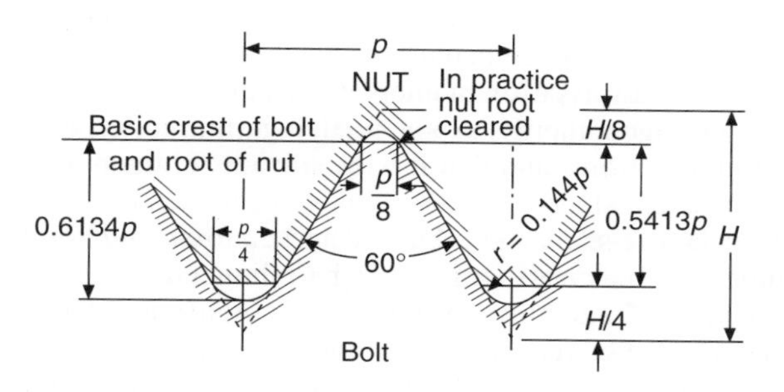

Fig. I.2 ISO Metric Thread

screw threads of triangular form (V-thread) with an included angle of 60°, available in coarse and fine pitches. It was adopted internationally and in the UK in 1965. See Fig. I.2.

isosceles triangle (*maths*) See: *triangle.*

isothermal process (*sci*) Any thermodynamic process carried out at a constant temperature.

iteration (*maths*) Obtaining a result by repeatedly performing the same sequence of steps until a specified condition is satisfied. Some equations can only be solved graphically or by methods of successive approximation of the roots. These are called *iterative methods*. The two methods of successive approximation are the *algebraic method* and the *Newton–Raphson formula*. Both methods rely on a reasonably good first estimate of the roots being made. The Newton–Raphson formula (often called Newton's method) may be stated as follows: If r_1 is an approximate value of the root of the equation $f(x) = 0$, then a closer approximation to the root (r_2) is given by the expression:

$$r_2 = r_1 - \frac{f(r_1)}{f'(r_1)}$$

where $f'(x)$ is the first derivative of $f(x)$. The advantage of the Newton–Raphson formula over the algebraic method is that it can be used for any type of mathematical equation including ones containing trigonometric, exponential, logarithmic, hyperbolic and algebraic functions, and that it is usually easier to apply than the algebraic method.

Izod impact test (*matls*) An impact test used to determine the toughness of materials. It is similar to the Charpy test in that it uses a notched specimen that is struck by a controlled impact. However in the Izod test, the specimen is supported as a cantilever as shown in Fig. I.3. The energy absorbed in breaking or bending the specimen is a measure of its toughness.

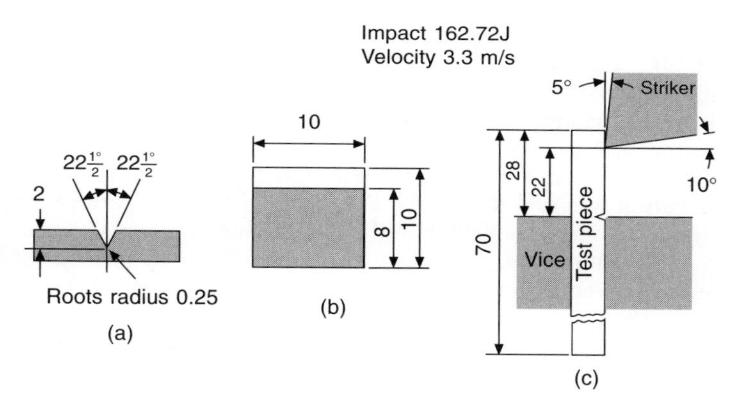

Fig. I.3 Izod Test (all dimensions in millimetres): (a) detail of notch; (b) section of test piece (at notch); (c) position of striker

J

jack (*mech*) Device for raising heavy loads off the ground by means of a ratchet system, a screw and nut system, or a hydraulic system. A simple example is shown in Fig. J.1.

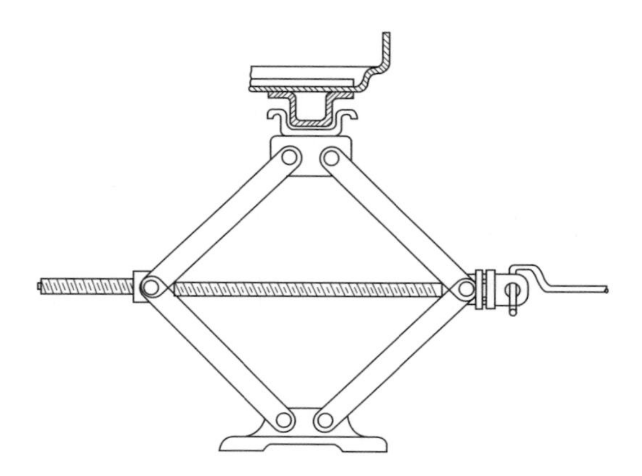

Fig. J.1 Simple Screw and Nut Scissor Type Jack

jib crane (*mech*) Crane for supporting heavy machinery, comprising a post, a tie and a jib, as shown in Fig. J.2. A common example of a triangle of forces as a vector diagram in text books dealing with engineering science and mechanics. See: *triangle of forces*.

joule (*sci*) SI unit of energy or work, symbol J; it is the work done when 1 newton of force moves 1 metre in the direction of

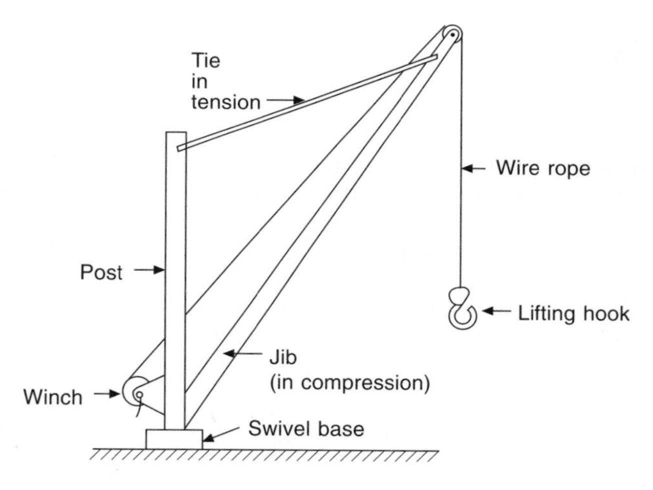

Fig. J.2 Jib Crane

the force. In electrical engineering 1 kWh = 3.6×10^6 J. The joule is also a unit of heat energy where 4186 J are required to raise 1 kg of pure water through 1°C.

journal bearing (*mech*) A bearing consisting of a cylindrical shaft or journal supported in a lubricated sleeve of bearing material, for example the bottom end bearings of internal combustion engines.

Karnough map (*maths*) A technique for simplifying truth tables. The Boolean expressions are depicted on a Karnough map so that any cells on the map having a common vertical side or a common horizontal side are grouped together to form a *couple*. The simplified Boolean expression for a couple is given by those variables common to all cells in the couple. An example of the simplification of a two-variable Boolean expression is shown in Fig. K.1.

Inputs		Output Z	Boolean Expression
A	B		
0	0	0	$\bar{A} \cdot B$
0	1	0	$\bar{A} \cdot B$
1	0	1	$A \cdot \bar{B}$
1	1	0	$A \cdot B$

Truth table

B \ A	0 ($\bar{A}$)	1 (A)
0($\bar{B}$)	$\bar{A} \cdot \bar{B}$	$A \cdot \bar{B}$
1(B)	$\bar{A} \cdot B$	$A \cdot B$

Matrix

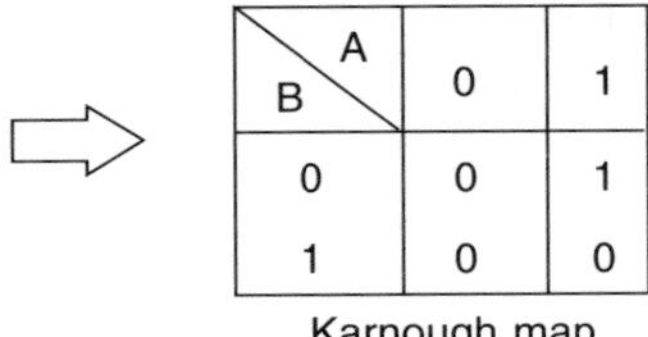

B \ A	0	1
0	0	1
1	0	0

Karnough map

Fig. K.1 Two-variable Karnough Map

kelvin (*sci*) SI unit of temperature used in many scientific and thermodynamic calculations, symbol K. (Note that the degree sign ° is omitted.) The magnitude of one unit on the kelvin scale is equal to one unit on the celsius scale. However, whilst 0°C is the freezing point of water, 0 K is *absolute zero* (–273.16°C). Thus, for example, 10°C = 10 + 273.16 K = 283.16 K.

Kelvin's coupling (*mech*) The classical solution to prevent movement within the six degrees of freedom without duplicated or redundant restraints, as shown in Fig. K.2. The tripod of supporting balls rest in a trihedral hole, a V-groove and a plain recess. Provided the resultant force of gravity and any extraneous

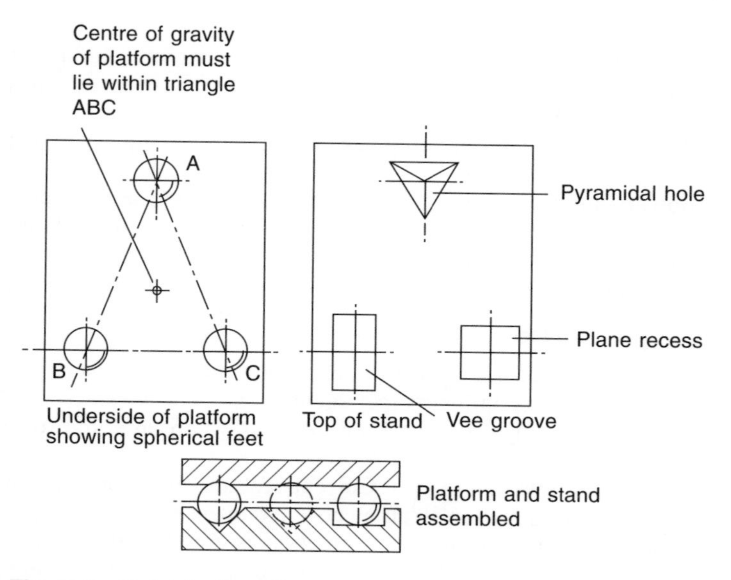

Fig. K.2 Kelvin's Coupling

forces act within a triangle joining the centres of the balls, the following conditions will be satisfied:

- the plates are mutually restrained;
- the plates are free to expand or contract without distortion and without affecting the integrity of the coupling;
- the top plate is supported without distortion.

key (*mech*) A component used to transmit rotary motion (torque) between a wheel and a shaft in a positive manner. The key is a piece of rectangular or square steel that fits into a groove in the shaft, parallel to its axis, and a corresponding groove in the wheel (half its thickness in each). See: *Feather key*; *gib-head key*; *Woodruff key*.

keyboard (*comp*) Group of switches for data input into a computer, usually arranged in an alpha-numeric matrix and based on a typewriter (QWERTY) key layout.

kilobyte (*elec*) Unit of computer memory storage equal to 1024 *bytes*.

kilowatt-hour (*elec*) Unit of electrical energy used commercially. It is equivalent to 1000 watts of power consumed for 1 hour, symbol kWh. See: *joule*.

kinetic energy (*sci*) Energy due to motion. A common definition of kinetic energy is the work done by a moving body when brought to rest.

Kinetic energy of a body with linear motion = $^1/_2mv^2$ where v is the linear velocity in metres per second.

Kinetic energy of a rotating body = $^1/_2I\omega^2$ where ω is the rotational velocity in radians per second.

Kirchhoff's laws (*elec*) (1) Current law: the algebraic sum of currents at any junction in an electric circuit is equal to zero, or the sum of currents entering a junction is equal to the sum of currents leaving that junction. (2) Voltage law: the algebraic sum of the e.m.f.s in an electric circuit is equal to the algebraic sum of the potential differences taken in order around the circuit.

knife-edge supports (*mech*) **1** An idealized, frictionless support used when considering beam theory to simplify calculations, the implication being that all the reaction forces are concentrated at precise points. **2** Support system affording minimum friction used in static balancing equipment.

knuckle press (*mech*) A mechanical power press used for forging, with a knuckle (toggle) joint mechanism that magnifies the force applied by the crank and connecting rod. This gives a short stroke to the ram but provides great pressure with a slow heavy squeeze, allowing time for the metal to flow to the required shape.

knuckle thread (*mech*) A screw thread profile with a rounded square section with a radius of 1/4 the pitch; used for some rough applications such as heavy duty hose connections.

knurling (*mech*) Process of flow-forming a series of overlapping, fine left- and right-hand grooves in a cylindrical material, for ease of grip, through a *burring* action. This is achieved by forcing a pair of serrated-patterned rollers, known as a knurling tool, into the surface of the workpiece whilst it is rotating in a lathe.

lag (*sci*) Time delay between the input to a system and the required output.

laminar flow (*sci*) Flow regime of a liquid through a pipe, where all the particles at a given radius from the pipe centre move at the same velocity with no turbulence, the velocity being highest at the centre and lowest at the pipe walls. Laminar flow occurs at low *Reynolds numbers*, low velocities, high viscosities and low densities. See: *Reynold's number*.

latent heat (*sci*) Energy absorbed or released when a material changes state at a constant temperature. The *latent heat of vaporization* is the energy required to change a liquid into a gas at the boiling point without a change in temperature. The *latent heat of fusion* is the energy required to change a solid into a liquid at the melting point without a change in temperature.

lathe (*mech*) See: *automatic lathe*; *capstan lathe*; *centre lathe*; *CNC lathe; turret lathe*.

LCM Abb. (*maths*) Lowest common multiple. The smallest number which is exactly divisible by each of two or more numbers is called the lowest common multiple. For example, consider the LCM for the numbers 12, 30, 42. The smallest single number into which these numbers will exactly divide is 420. This is their lowest common multiple. See: *multiple*.

lead (*matls*) Heavy, corrosion-resistant metal (melting point 327.4°C), used for lining chemical vats, it is also the basis of lead–tin alloys used for white metal bearings and soft-solders, also used as an alloying element in free-cutting steels.

lead (*mech*) The axial distance moved by a nut along a bolt in one revolution, where lead = pitch × number of starts. See: *pitch*.

'lead-through' programming (*CAE*) A method of programming industrial robots in which a skilled craftsperson holds the tool in the end effector and carries out the work that is to be done eventually by the robot. The robot computer memorizes all the moves and can repeat them exactly over and over again as many times as is required.
Lenz's law (*elec*) The polarity of an induced e.m.f. always opposes the change that produced it. This is considered by some to be analogous to Newton's third law of motion: for every action there is an equal and opposite reaction.
letter address system (*CAE*) See: *word address system.*
lever (*sci*) A simple machine that applies the *principle of moments* to overcoming a load by means of an effort force. There are three orders of lever and these are shown in Fig. L.1.
lift pump (*sci*) The term used to refer to any pump used to raise a liquid from one level to a higher level, e.g. a pump used for raising water from a well to a storage cistern. See: *force pump.*
light-emitting diode (LED) (*elec*) A semiconductor diode which emits light as the result of an electroluminescent effect as electrons and 'holes' combine at the junction. The light emitted is of specific colours that may be red, orange, yellow, green or blue; infrared is also emitted. See: *diode.*
limit gauge (*mech*) A gauge used in production work to check if a component dimension lies between its specified (upper and lower) limits of size. There are usually two gauge elements, a GO element into which the component should fit if it is not over-size, and a NOT GO element into which the component should not fit unless it is undersize. A limit gauge is not a measuring device, and can only determine whether the component can be *accepted or rejected.* It cannot indicate the actual size.
limit of proportionality (*sci*) The point on the stress-strain curve of the results of a tensile test on a material where strain ceases to be proportional to stress producing it. For any given

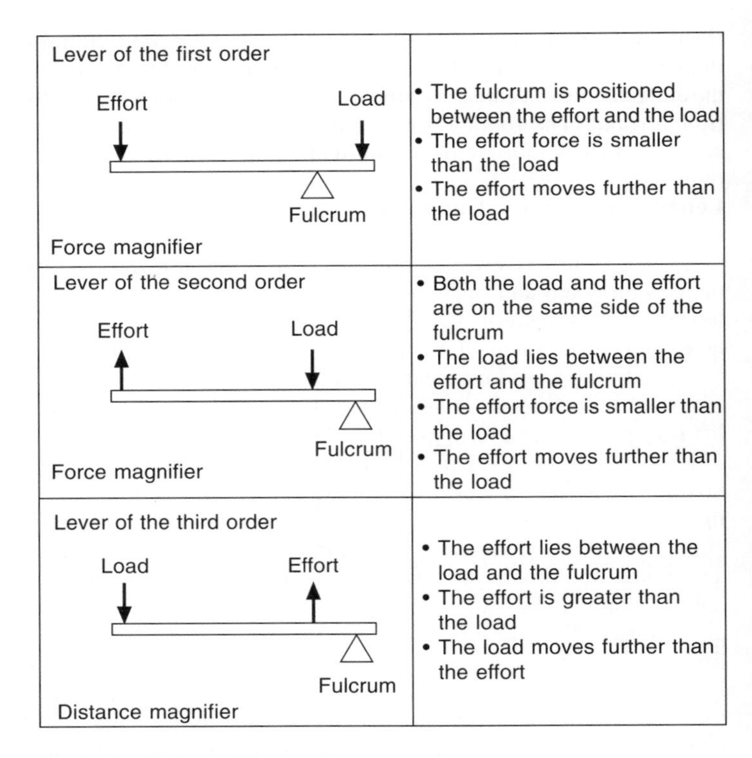

Fig. L.1 Orders of Levers

material it is the maximum tensile stress for which Hooke's law is applicable. See: *Hooke's law*; *elastic limit*; *Young's modulus*.
linear correlation (*maths*) When a straight line can be drawn through all the points plotted on a graph, *perfect linear correlation*

is said to exist. If the straight line has a positive gradient, *positive or direct linear correlation* exists. If the straight line has a negative gradient, *negative or indirect linear correlation* exists. When there is no apparent association between the points on a graph (random scatter) no correlation exists. The degree of correlation between two variables is expressed by a coefficient of correlation (r). This can be calculated by the product-moment formula:

$$r = \frac{\sum xy}{\sqrt{\{(\sum x^2)(\sum y^2)\}}}$$

linear equation (*maths*) Any equation of the first order, i.e. any equation where the unknown is raised to the power of 1. Such an equation gets its name from the fact that it will always give a straight-line graph. A linear equation is written as $y = ax + b$ or, in some older texts, $y = mx + c$. In both examples y is the dependent variable and x is the independent variable, a or m is the *coefficient* of the independent variable and controls the gradient of the graph, and b or c is a *constant* that controls the *intercept* or point where the straight line cuts the y-axis. The coefficient is the ratio $\delta y/\delta x$.

linear regression (*maths*) See: *regression.*

linear variable differential transformer (LVDT) (*mech*) A transducer used for linear measuring devices that employ electronic magnification of the deflection of the measuring stylus. The principle is shown in Fig. L.2. Any deflection of the stylus that moves the soft-iron core from its equilibrium position changes the magnetic flux pattern. This results in the induced current in one secondary winding becoming greater or smaller at the expense of the induced current in the other secondary winding, thus providing a signal that can be processed to show the magnitude of the deflection on an analogue or digital read-out. The primary winding is fed from a frequency-stabilized, high-frequency oscillator in

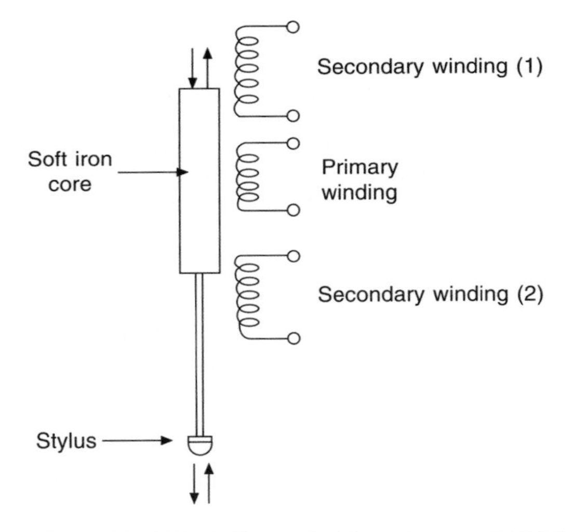

Fig. L.2 Linear Variable Differential Transformer (LVDT)

order to keep the physical size of the transducer to a minimum.
liquid (*sci*) Phase of matter between the solid state and the gaseous state in which the molecules of a substance are sufficiently mobile to allow that substance to adopt the shape of a container and retain an almost fixed volume.
liquid crystal display (LCD) (*elec*) A visual display in which a thin layer of liquid crystal is contained between two electrodes. The application of an electrical potential across the electrodes changes the light transmission properties of the liquid crystal material.
load cell (*sci*) Sensing element of an instrumentation system that converts force into a more suitable electrical or hydraulic analogue for remote display or recording, e.g. an electrical strain

gauge where there is a change in electrical resistance and, therefore, current flow when the cell deforms under load.

logarithm (*maths*) The power to which a number is raised to give another number, e.g. $x^n = y$, where n is the logarithm and x is called the base, written as $n = \log_x y$, meaning n is the log to the base x of y. Logarithms were often used to assist in large calculations before computers and calculators became common. Natural logarithms occur frequently in engineering and have a base of 2.71828... represented by 'e', e.g. $\log_e y$. This is often written as $\ln y$. Natural logarithms are also called *hyperbolic* logarithms or *Naperian* logarithms. Common logarithms have a base of 10, commonly written $\log y$ without the base being stated.

logic circuits (*elec/comp*) Circuits using discrete state (binary) signals that can be expressed in terms of Boolean algebra rather than continuous (analogue) state signals. *Combinational logic* functions have outputs that depend on the current inputs, and the functions AND, OR and NOT form the building blocks of complex circuits. *Sequential logic functions* have outputs that depend also on previous states of the system.

logic gate (*elec*) A logic gate is a circuit that may have a number of inputs but only one output which will be at logic 1 or logic 0 depending upon the inputs. A simple example is shown in Fig. L.3. This is an OR gate because if A *or* B *or* C is at logic 1 the output will be a logic 1 and the lamp will light.

logic levels (*elec*) Low logic levels are referred to as logic 0 and high logic levels are referred to as logic 1. Consider a flash lamp: when the switch is open no current flows and the voltage across the bulb is zero. This is equivalent to logic 0. When the switch is closed a current flows and the potential across the bulb is approximately +3V. This is logic 1 and it is called *positive* logic. If logic 0 is –5 V and logic 1 is 0 V it is still *positive* logic. However, if logic 0 is +3 V and logic 1 is 0 V it is called *negative* logic. Note: in practice potentials are rarely exact so that logic

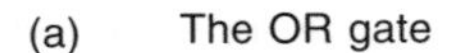

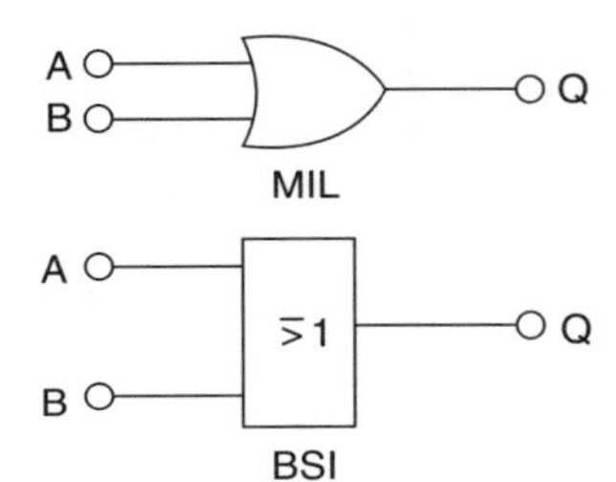

(b) OR gate truth table

A	B	Q
0	0	0
0	1	1
1	0	1
1	1	1

Fig. L.3 OR Gate and Truth Table

levels are represented by bands of voltage. For example, logic 0 may be 0 to 1 V and logic 1 may be 3 to 4 V.

longitudinal (*sci*) Lengthways. For example, longitudinal measurement is the measurement of length.

loop (*CAE*) See: *nested loop*.

lost wax process (*mech*) See: *investment casting*.

loudspeaker (*elec*) A transducer that converts electrical signals into audible sound signals.

lubricant (*sci*) Substance used to keep bearing materials apart to reduce friction and wear. Natural and synthetic oils are commonly used in engineering and these also act as a coolant and have good thermal and viscosity properties; additives are used to control sludge formation and the pH value. Grease is also used as it can adhere to surfaces that can only be lubricated on an intermittent basis and it also keeps dirt out. Solids such as graphite are useful at higher temperatures. Compressed air may also be used in certain circumstances to keep bearing surfaces apart.

lubrication (*sci*) See: *boundary lubrication*; *fluid lubrication*.

luminous intensity (*sci*) Measure of the light-emitting property of a light source, quantity symbol I_v, measured in the SI system in candela, symbol cd. The luminous intensity of a monochromatic light source = 1 cd if the radiant intensity = 1/683 watt per steradian in a given direction when the frequency of the emission is 540×10^{12} Hz. See: steradian.

M

machine (*sci*) Any device or system used for transferring energy in order to overcome a load or resistance at one point by the application of an effort force at another point. Generally, any device used for the conversion of motion. Typical examples of simple machines are: pulley blocks, screw jacks, levers or crowbars.
machine code (*comp*) Basic instruction code used for programming in a form that a microprocessor can directly understand, consisting of a series of bytes stored in the memory, each interpreted by the microprocessor as either an instruction or data. Programming in machine code is difficult, tedious and prone to error.
machine tool (*mech*) Any power-driven machine designed for shaping and cutting metal, e.g. a centre lathe, milling machine, etc. The purpose of a machine tool is to drive and guide a suitable cutting tool. The cutting tool is selected, installed and set to suit the job in hand. Increasingly, machine tools are computer programmable and are capable of automatic operation, allowing a precise sequence of operations to be followed repeatedly for manufacturing large numbers of components.
Maclaurin's series (*maths*) Also known as Maclaurin's theorem. It is a simplification of Taylor's series, putting $a = 0$ and can be expressed as:

$$\mathrm{f}(x) = \mathrm{f}(0) + x\mathrm{f}(0) + \frac{x^2}{2!}\mathrm{f}(0) + \ldots$$

See: *Taylor's series.*
macro-language (*CAE*) Most CNC systems have a macro-

programming language. Fanuc 0M controllers have two languages: Macro A or Macro B. The macro-language allows the user to store variables within registers (e.g. #100) and use mathematical functions such as sine, cosine, tangent, square root, etc. The macro-language is like any other conventional computer language such as BASIC, and can be used for such applications as:

- Positioning holes on a pitch circle.
- Elliptical path machining.
- In-cycle gauging.
- Setting up tools and workpieces using probes.

magnetic circuit (*elec*) Refers to all the space occupied by magnetic flux, and can be enclosed using an annular iron core providing a path for the lines of flux and resulting in a uniform field.
magnetic field (*elec*) Force field produced by permanent magnet or electromagnet, represented graphically by lines of flux which form concentric closed loops reducing in strength with distance from the magnet and which, when stretched, try to return to their natural shape. Lines of flux are assumed to flow from the north pole to the south pole of the magnet in space and from the south to the north pole inside the magnet. Like magnetic poles repel and unlike poles attract.
magnetic flux (*elec*) Lines of force representing a *magnetic flux field*, quantity symbol, Φ, unit the *weber*, Wb (pronounced *vayber*).
magnetic flux density (*elec*) Measure of the magnetic flux density per unit area of the cross-section of a magnetic field, quantity symbol, B, unit the tesla, T, calculated from:

$$B = \Phi/A$$

where A is the cross-sectional area of the magnetic field; 1 T = 1 Wb/m^2.
magnetic materials (*matls*) Iron, nickel, cobalt and gadolinium

are the only pure metals that can be magnetized and of these iron is outstanding. They are all referred to as *ferromagnetic* materials. Alloys of these metals together with aluminium and/or tungsten are used to make very powerful permanent magnets.

magnetic tape (*CAE*) This was introduced as a compact means of loading complex programs into CNC machines and saving programs for reuse when a subsequent batch of components was required. It was widely used in conjunction with computer-assisted programming. Nowadays, magnetic (floppy) disks are much more likely to be used. See: *manual data input*, *punched tape input*; *conversational input*; *direct numerical control*.

magnetomotive force (m.m.f.) The force that causes the magnetic flux of a magnetic field, symbol F, analogous to *e.m.f.* causing current flow in an electric circuit. The units of m.m.f. are derived from electromagnetism: the flux is proportional to the number of turns of the coil and to the current through the coil; hence mmf is commonly measured in ampere-turns (At) although the strict SI unit is the *ampere*.

malleability (*matls*) The property of a material that enables it to withstand being flow-formed to shape by the application of compressive forces as in forging and rolling. (See: *ductility*)

management word (*CAE*) These are words used in CNC programming that are not related to dimensions, i.e any word commencing with N, G, F, S, T, M, or any word in which these characters are implied. For example, N is used for block numbers, G is used for a preparatory code or function, F is a feed-rate command, S is a spindle-speed command, T is a tool number and M is a miscellaneous command such as turning the coolant on or off. See: *dimensional word*.

mandrel (*mech*) A work-holding device consisting of an accurately ground, case-hardened, cylindrical steel bar. The work is supported on the mandrel and the mandrel, in turn, is supported between the lathe centres. Mandrels are used for the support of work that has been drilled and reamed so that the external diameter can be turned concentric with the reamed hole. The mandrel, which has a slight taper, is pressed into the workpiece with a mandrel press. The work may then be machined, the friction between the mandrel and work resisting the cutting force. For this reason the work must be mounted so that the direction of the cutting forces tends to drive the work towards the large end of the taper, thus increasing the wedging action. Various sizes and lengths of mandrels are available so that the small end fits through the hole but the larger end does not.

mandrel press (*mech*) A small, bench-mounted, manually operated press in which the effort magnification is achieved by a rack and pinion, the pinion being rotated by the operating lever. As its name implies, a mandrel press is used for pressing a tapered mandrel into a hollow component so that it can be held between centres for turning external diameters. See: *mandrel*.

manometer (*sci*) Instrument for measuring pressure differential, consisting of a transparent U-tube with one leg connected to a system containing a fluid whose pressure is to be measured, and the other open to the atmosphere. The U-tube contains a liquid such as water for low pressures or mercury for higher pressures. With both legs of the U-tube subjected to atmospheric pressure the levels of the liquid will be the same in both legs. If the system leg is subjected to a pressure higher than atmospheric pressure the level of the liquid will rise in the open leg. If the system leg is subjected to a pressure lower than atmospheric pressure (partial vacuum) the level of the liquid will fall in the open leg. The difference in levels is proportional to the pressure of the fluid being measured relative to atmospheric pressure, as indicated on a scale located between the legs.

manual data input (*CAE*) This enables the operator to enter complete programs, edit programs and make adjustments whilst setting the machine by pressing buttons on the control unit. Little

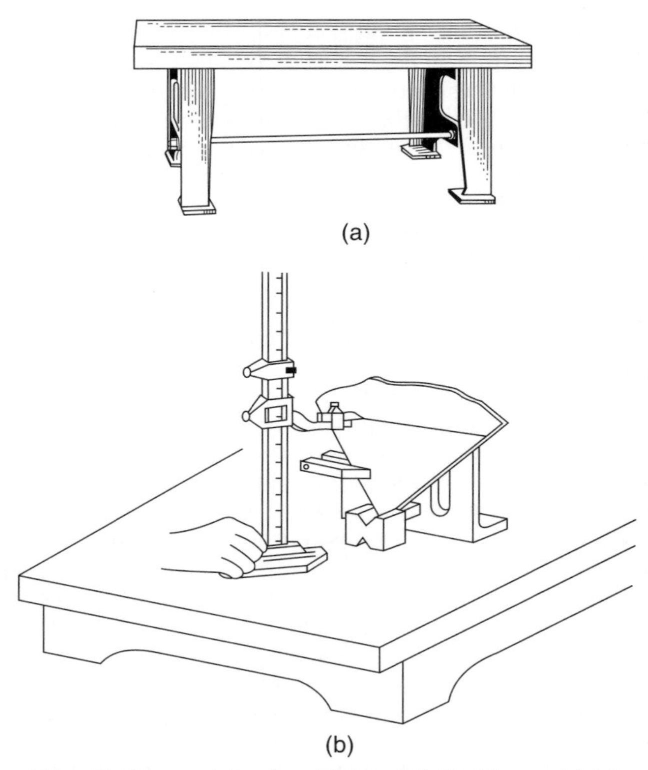

Fig. M.1 Marking-out (surface) table: (a) Marking-out table (datum surface); (b) use of a datum surface when marking out [WPPM, Figs 3.6(a) and 3.10(b)].

used for loading complete programs nowadays as it increases the machine's 'idle-time'. See: *conversational data input*; *punched tape input*; *magnetic tape input*; *direct numerical control*.

marking out (*mech*) The process of marking lines on a workpiece using such instruments as a scriber, odd-legs, rule and dividers, according to the design dimensions, prior to cutting and shaping. The lines are scribed relative to a datum line (such as a centre line) or a datum edge. The surface to be marked out is generally coated with a spray-on, quick-drying marking ink to make the lines show up clearly. See: *datum line*, *datum edge*; *datum surface*; *vernier height gauge*; *witness marks*.
marking-out table (*mech*) Sturdy table made of cast iron with a flat top and square edges, used to provide a *datum surface* for precisely marking out work material prior to machining, for example as shown in Fig. M.1.
martensite (*matls*) The product of a non-equilibrium phase transformation that occurs when steel is cooled too quickly for the normal equilibrium transformations of the iron–carbon diagram to occur. Martensite can be most conveniently considered as a *supersaturated* solid solution of carbon in body-centred cubic iron, as opposed to austenite that is a simple solid solution of carbon in face-centred cubic iron. Unlike austenite which is soft and ductile, martensite is the hardest and most brittle of the phases encountered in steel. Under the microscope martensite appears as a uniform mass of needle-shaped crystals.
mass (*sci*) Measure of the quantity of matter that can be defined in various ways in engineering; quantity symbol m, SI unit is the kilogram (kg). Mass is commonly considered to be a measure of a body's inertia or resistance to change in motion and is a constant property wherever measured. Mass is often confused with *weight* in everyday language as 'weight' is often expressed using units of mass. Mass can be defined in terms of gravitational force in accordance with *Newton's law of gravity*, and also according to Einstein's special theory of relativity where mass is considered to be a measure of the total energy content. See: *weight*.
mathematical model (*maths*) A mathematical expression of

events to predict future outcome. Such a model is usually only accurate for a particular condition and range. Dynamic models link past with present behaviour and are used in engineering for control systems or other purposes such as fault diagnosis, by comparing current behaviour with the expected behaviour of the model. Dynamic model development from plant measurements is known as identification. Compare mathematical models with non-mathematical models such as *fuzzy models* and *neural networks*.

matrix (*maths*) Set of values arranged in a rectangular array, allowing mathematical operations to be performed on them. Multivariate data sampled from plants are usually arranged in a matrix in vertical columns for analysis purposes.

Mazac (*matls*) An alloy of high-purity zinc containing up to 4% aluminium and 2.7% copper specially developed and widely used for the production of small components by the pressure die-casting process. See: *die-casting*.

mean (*stats*) The arithmetic mean (or average) of a set of n values is the sum of the values divided by n, and is used as a measure of the central tendency of a process.

mean effective pressure (*sci*) A performance criterion of any reciprocating internal combustion engine; it is the pressure that would act on a piston during one stroke if the pressure were to remain constant throughout the stroke and produce the same net work output.

mechanical advantage (*sci*) Ratio of the load to the effort for a machine when work is being done.

mechanics (*sci*) The study of forces acting on bodies. *Statics* is the study of the action of forces on stationary bodies, e.g. frameworks. *Dynamics* is the study of the action of forces that create a change in motion of a body. Kinematics is the study of the geometry of motion. Fluid mechanics is the study of the relationship between forces and fluids at rest and in motion.

median (*maths*) The middle value of a series of numbers. For example the median of 1, 2, 3, 4 and 5 is 3, since two numbers lie either side of three. Therefore there can only be a real median of an odd number of members that are *ranked* in ascending order. If there is an even number of members, the median is the mean of the central two numbers.

megabyte Unit of computer storage capacity equal to 1024 kilobytes.

member (*stats*) An individual value within a set of statistical data.

mensuration (*maths*) That branch of mathematics concerned with the measurement of lengths, areas and volumes.

metal (*matls*) The majority of the naturally occurring elements in the periodic table in which the atoms are held together by a *metallic bond*. It is this bond that give metals their distinctive properties and distinguishes them from the non-metals. The characteristic properties of metals include high electrical and thermal conductivity, high strength and high density when compared with non-metals. When polished, metals have highly reflective surfaces. With the exception of mercury, all metals are solids at room temperature. Metals may be subdivided into the alkali metals, the alkaline rare-earth metals, the transitions metals (including those widely used in engineering), the actinides and the lanthanides.

metal-cutting tool angles (*mech*) All metal cutting tools are essentially wedge shaped. An example is shown in Fig. M.2. It can be seen that the three angle that make up the metal-cutting wedge are the rake angle which controls the chip formation, the clearance angle that prevents the tool from rubbing and enables it to penetrate the surface of the work, and the wedge angle itself. Since the three angles add up to a right-angle, any change in one angle affects the remaining angles. Generally the clearance angle is fixed at between 5° and 7°. Increasing the rake angle increase the cutting efficiency but reduces the wedge angle and weakens the tool point. Increasing the wedge angle strengthens the tool

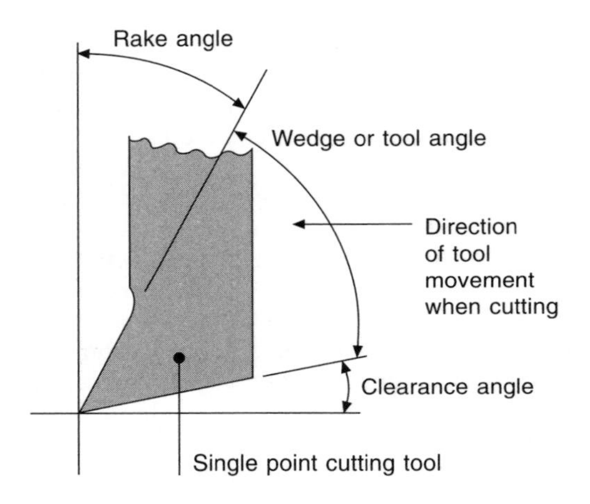

Fig. M.2 Metal Cutting Tool Angles

point but reduces its cutting efficiency. The choice of wedge angle and rake angle is always a compromise for any given application. See: *clearance angle, rake angle.*

metastable (*elec*) A state that is not truly stable but only apparently stable, often due to the slowness in attaining equilibrium conditions.

method of sections (*sci*) A technique used in the study of frameworks and structures, where the force that acts in one particular member is required rather than those of all the structural members. The framework is imagined to be split into two parts, the dividing line cutting the member under investigation, and usually two others, if the framework is comprised of triangles, as most are. The part of the framework to one side of the line will only stay in equilibrium if additional forces are now added. These additional forces are equal to those normally provided by the rest of the structure. The value of the unknown force of the member under investigation

can be found by taking moments about the intersection of the two forces of the other two divided members. The other two forces will have a zero moment about their intersection, usually leaving one unknown value in a moment equation.
metre (*sci*) SI unit of length, unit symbol m. See: *Appendix 4*.
metric system (*sci*) Decimal system of units developed in the eighteenth century based on the metre, gram and second. This later evolved into a system based on the metre, kilogram and second which became the Système International d'Unités (International System of Units) or SI system for short.
micrometer (*mech*) Instrument for measuring the length of components in millimetres to two decimal places; it comprises a U-shaped frame, extending spindle that is adjusted by a calibrated barrel and thimble, and an anvil at the other end. The bore of the barrel is threaded $^1/_2$ mm pitch, the surface is graduated in $^1/_2$ mm divisions and the rim of the barrel is subdivided into 50 divisions. The spindle is adjusted with the thimble close to the anvil and spindle around the component, and a ratchet is fitted on the thimble end to ensure that a consistent force is used. The length is then read from the scale on the barrel to 0.5 mm and then to 0.01 mm on the thimble, as shown in Fig. M.3. The usual range is over 25 mm, with successively larger instruments being 25–50 mm, 50–75 mm, and so on. Inside micrometers are available and are used to measure the bore of holes and other inside linear dimensions.
microphone (*elec*) A transducer that converts sound energy into electrical energy that is an analogue of the original sound.
microprocessor (*comp*) The central processing unit of a micro-computer to which RAM, ROM and ports are added. It contains *registers*, control unit, arithmetic and logic unit, manufactured from an integrated circuit. See: *CPU*.
microwaves (*elec*) Electromagnetic waves whose wavelengths range from 0.001 m to 0.3 m (frequency 300 GHz to 1 GHz).
mid-ordinate rule (*maths*) With reference to Fig. M.4, the

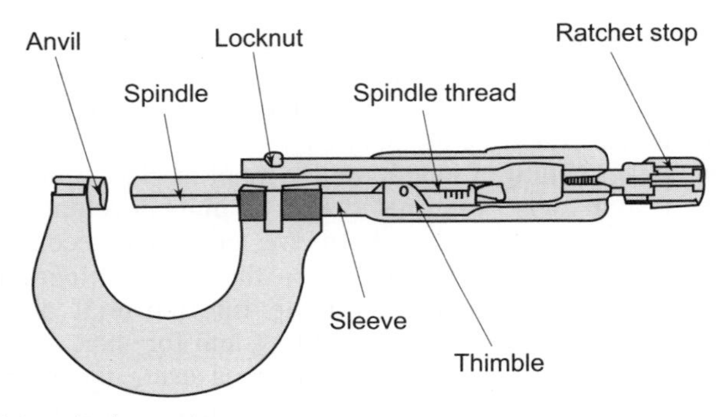

(a) External micrometer

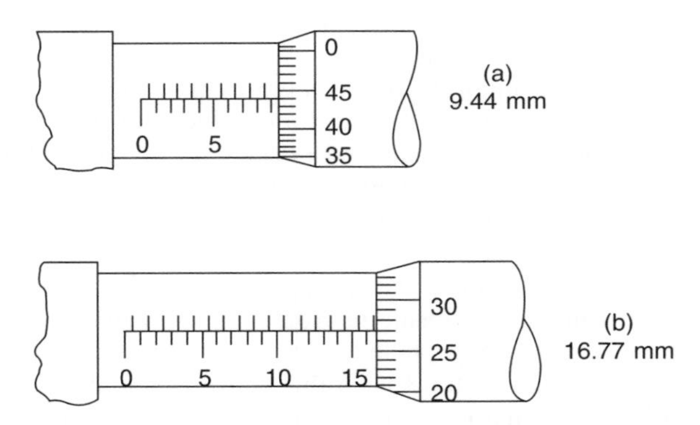

(b) Metric micrometer readings examples

Fig. M.3 External micrometer

approximate area under the curve AB can be determined by dividing the base OX into any number of equal intervals (x) (the greater the number of intervals, the greater the accuracy), with ordinates

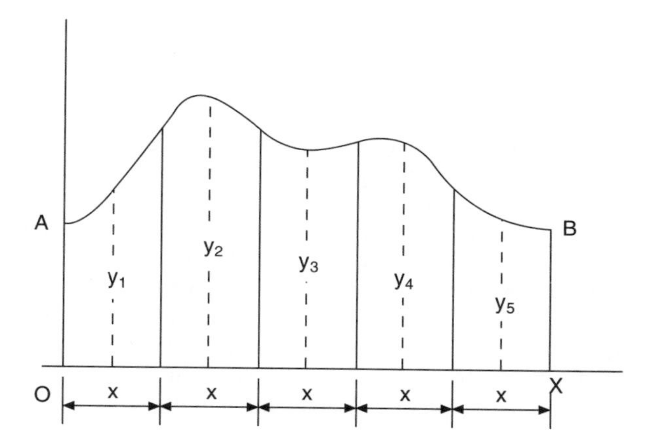

Fig. M.4 Mid-ordinate Rule

erected at the centre of each interval as shown by the broken lines. The ordinates y_1, y_2, y_3, y_4 and y_5 are measured and the area of the figure determined as follows:

$$\text{area} = x(y_1 + y_2 + y_3 + y_4 + y_5)$$

Expressed generally,

$$\text{area} = (\text{width of interval}) \times (\text{sum of the ordinates}).$$

MIG welding process (*mech*) The *metal inert gas* process is an electric arc-welding process using a wire electrode filler rod fed automatically from a coil of uncoated wire, the arc being shielded against oxidation by use of an inert gas such as carbon dioxide or argon. Therefore no flux is required and a clean and neat weld is produced. MIG welding is used extensively in the motor vehicle and aircraft industries.

mild steel (*matls*) These are plain carbon steels with a low carbon content ranging between 0.15% and 0.25%, and therefore they are largely ferritic (ferrite = weak solid solution of carbon in iron). Therefore like all materials that are solid solutions they are soft and ductile but lack the strength of medium- and high-carbon steels. They do not respond to heat treatment except for subcritical annealing after cold-working.

mill scale (*matls*) This is the dry corrosion of steel to form a heavy scale of iron oxide on the surface of the metal when it is being hot rolled or hot forged. The high temperatures involved preclude the presence of water so the more familiar red, hydroxide rust does not form. The mill scale is grey in colour. See: *rusting.*

milling cutters (*mech*) Rotating, multi-tooth cutters used in milling machines for producing a variety of surfaces that are parallel to, perpendicular to, or at an angle to the work table of the machine. Some cutters have a profile for cutting formed surfaces such as corner radii and gear teeth. A selection of cutters and the surfaces they produce are shown in Fig. M.5.

milling machines (*mech*) Powerful machine tools for producing plain surfaces using rotating multi-tooth cutters. A typical *horizontal milling machine* is shown in Fig. M.6(a). It is so called because the spindle to which the cutter arbor is attached has a horizontal axis. A typical vertical milling machine is shown in Fig. M.6(b). It is so called because the axis of the spindle on which the cutters are mounted is vertical.

mirror imaging (*CAE*) This is a programming facility available on some control systems for milling and drilling operations that allows either a whole program or a subroutine to have the signs of its coordinates selectively reversed, as shown in Fig. M.7. Although controllers vary considerably, the ISO code for mirror imaging is G28 and typical program lines could be as follows:

N100 G28 X reverses the x-coordinates only for the subsequent subroutine.

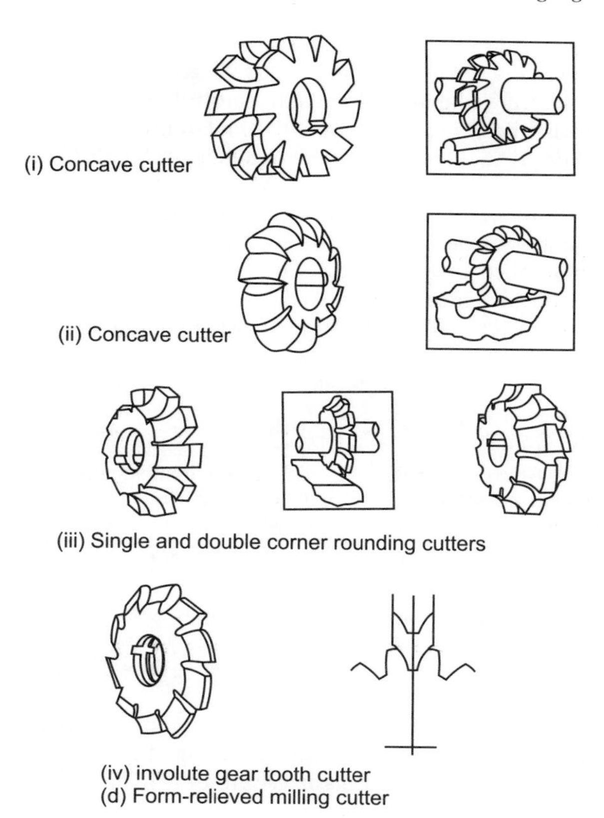

Fig. M.5(a) Horizontal milling machine cutters and the surfaces they produce

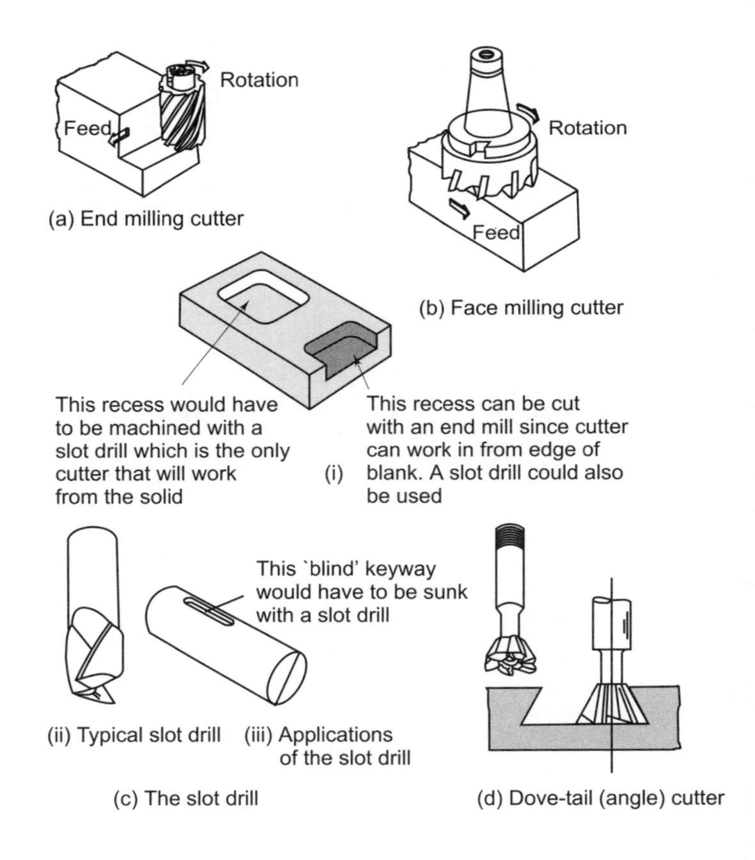

Fig. M.5(b) Vertical milling machine cutters and the surfaces they produce

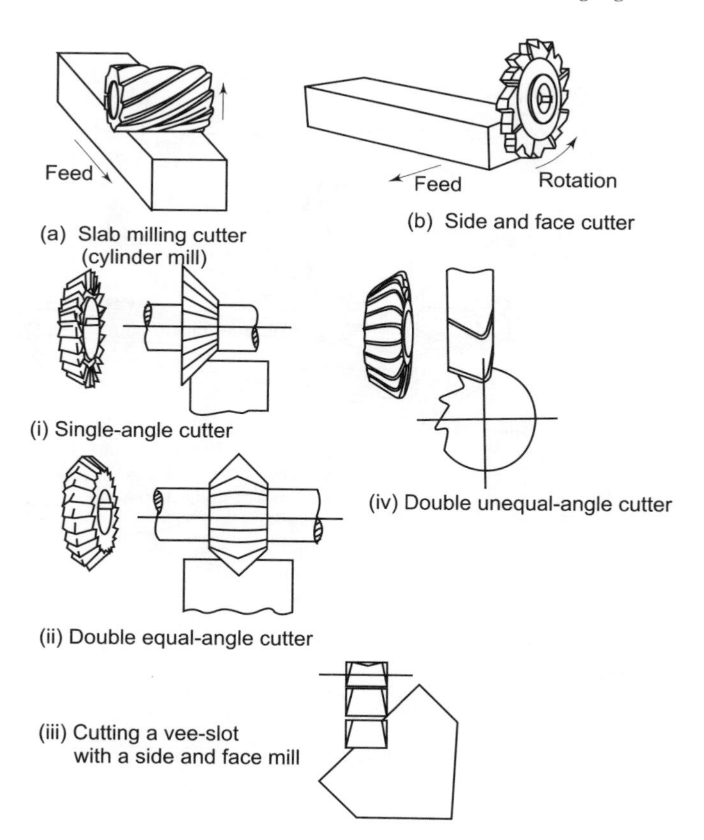

Fig. M.5(c) Angle Milling Cutter

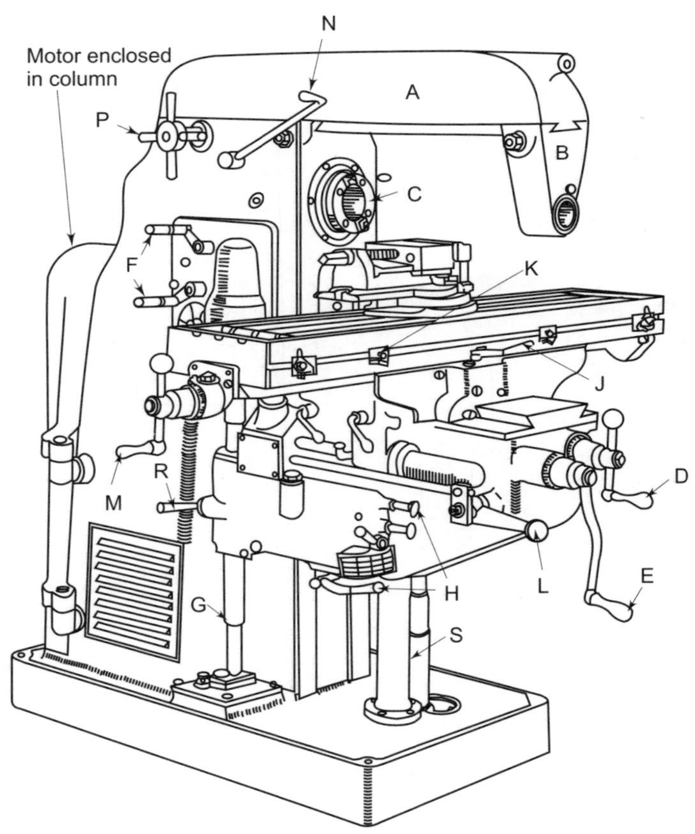

A, Overarm; B, Arbor supporting bracket; C, Spindle nose; D, Hand cross feed; E, Hand vertical feed; F, Speed change levers; G, Feed driving shaft (enclosed); H, Feed change levers; J, Table feed actuating lever; K, Feed trip; L, Rapid power feed control; M, Hand table feed; N, Starting lever; P, Wheel for moving overarm; R, Feed reversing lever; S, Tube to deliver cutting fluid to reservoir in base.

Fig. M.6(a) Horizontal Milling Machine

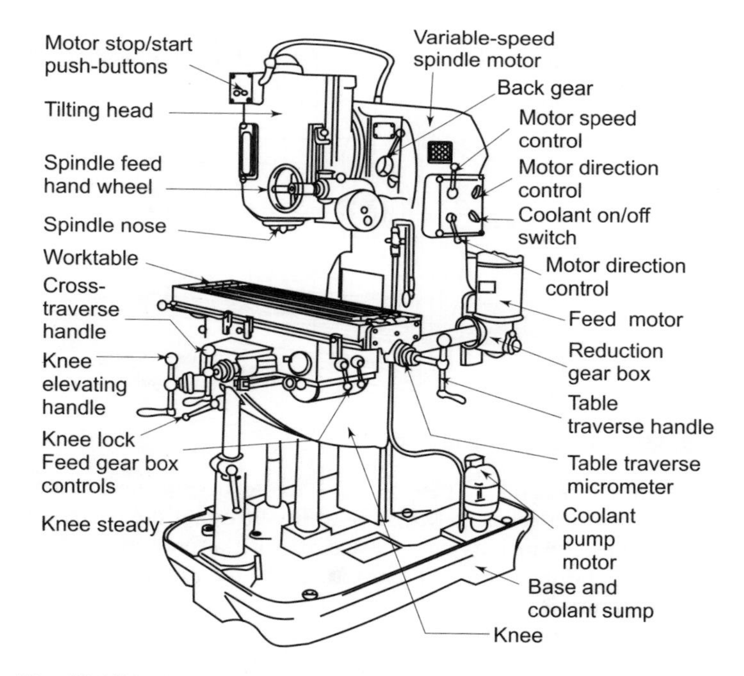

Fig. M.6(b) Vertical Milling Machines

N200 G28 Y reverses the *y*-coordinates only for the subsequent subroutine.

N300 G28 XY reverses both the *x*- and *y*-coordinates for the subsequent subroutine.

mixed number (*maths*) An integer followed by a proper fraction with which it is associated, e.g. $1^1/_2$.

modal function (*CAE*) Modal functions are preparatory commands that remain enabled (turned on) from block to block until they are

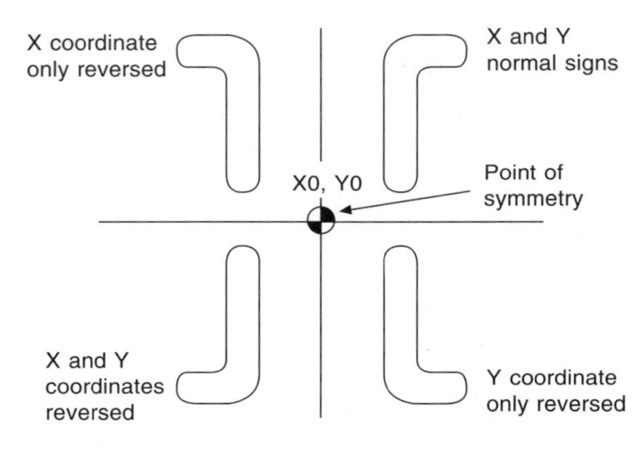

Fig. M.7 Mirror Imaging in CNC

disabled (turned off). They do not have to be repeated in each block. For example, using BS3635 codes, G90 establishes a program in absolute dimensions until G91 calls for the program to be in incremental dimensions. The program will then remain in incremental dimensions until G90 is used to change the program back into absolute dimensions. There are a large number of modal commands. However, always study the programming manual as different manufacturers do not follow the same standards. For example FANUK controllers use G20 in place of G90 and G21 in place of G91.

modal value (*stats*) Also called the 'mode', it is the most commonly occurring value in a set of numbers. If two values occur with the same frequency, the set is said to be bimodal.

modulus of elasticity (*sci*) Also known as Young's modulus of elasticity, it is the ratio of the stress applied (σ) to the strain produced (ε) for a material under direct stress, provided that the limit of proportionality is not exceeded and Hooke's law applies.

The quantity symbol is E, the units are the same as stress (N/m^2). Owing to the large magnitudes of the stresses involved (GN/m^2), the modulur is a measure of the material's resistance to tensile and compressive forces.

$$E = \frac{\text{stress}}{\text{strain}} = \frac{\sigma}{\varepsilon}$$

e.g. for steel $E = 210\ GN/m^2$.

modulus of rigidity (*sci*) The ratio of the shear stress applied to the strain produced provided the limit of proportionality is not exceeded and Hooke's law applies.

$$G = \frac{\text{stress}}{\text{strain}} = \frac{\tau}{\gamma}$$

e.g. for steel $G = 81\ GN/m^2$.

Mohs scale (*sci*) Early scale for evaluating the hardness of a material, consisting of a list of materials in order of hardness, the hardest being a diamond at 10. Mainly used for minerals rather than metals and compares resistance to abrasion. Any item in the list can scratch any item below it. Modern hardness tests tend to measure resistance to penetration.

moment (*sci*) The moment of a force is the turning effect of a force about a point, calculated from the product of the force multiplied by the distance from the axis perpendicular to the line of action of the force, units N m. In calculations, clockwise moments tend to be considered positive and anticlockwise negative. If there is no rotating action of a body then there must be a balance of moments, i.e. clockwise moments equal anticlockwise moments: this is the principle of moments. If the distance from the line of action to the fulcrum is multiplied once, this is referred to as the first moment; if the distance from the line of action to the fulcrum is multiplied twice, this is referred to as the second moment. Moments can be considered for other quantities besides force. For example, the moment of inertia, symbol I, units kg

m^2, is an important concept in rotational dynamics and is the resistance to rotational acceleration of a body, comparable to mass in linear dynamics; this is the second moment of mass about an axis, $I = mr^2$. See also: *first moment of area*; *second moment of area*.

moment of area (*sci*) See: *first moment of area*; *second moment of area*.

momentum (*sci*) In linear dynamics this is equal to the product of the mass and the velocity, mv, units kg m/s. In rotational dynamics this is equal to the product of the moment of inertia and the angular velocity in radians per second, $I\omega$, units kg m^2/s. Note that the units are different for linear momentum and angular momentum and so the two quantities cannot be directly added together.

Monostable circuit (*elec*) A circuit that is fully stable in one state but metastable in another state into which it can be driven for a fixed period of time by an input pulse. See: *bistable circuit.*

morse taper (*mech*) An international standard taper system that is self-securing. It is used for locating and driving a variety of machine tool cutters including twist drills and machine reamers. The size of the taper depends on the size of the drill or cutter diameter and details are available in tables of dimensions for different sizes. The taper has a tang at the end that fits into a slot in the machine spindle but the drive should be from frictional forces between the taper shank and the machine spindle. The taper shank is extracted from the spindle by driving a tapered drift through a hole in the machine spindle to force the tang downwards.

motor (*elec*) Machine for converting one form of energy into mechanical energy to do work. For example, an *electric motor* converts electrical energy into mechanical energy; an *internal combustion engine* converts the chemical energy of the fuel into mechanical energy.

mould (*mech*) **1** The impression of a shape to be cast in a foundry, made of sand or metal. **2** The metal dies used for moulding plastic (polymeric) materials to shape.

moving-coil meter (*elec*) Sensitive direct-current measuring instrument. The current to be measured is passed through a coil of fine copper wire wound on a light former suspended between poles of a permanent magnet creating a radial field. Jewelled bearings are used to minimize friction. The current flowing through the coil and the magnetic field produce a deflecting torque. Current is passed to the coil via two contrawound spiral springs which also provide the restraining torque. Damping is provided by eddy currents induced in the coil former. Also the meter may be adapted to form a voltmeter. This type of analogue meter has a uniform scale and is very accurate. Its range may be increased by the use of voltage multipliers and current shunts. See: *shunt*; *voltage multiplier*.

moving-iron meter (*elec*) (1) *Attraction type*: an analogue meter in which a vane attached to the pointer is drawn into a solenoid against the opposing force of a spring. The greater the current flowing through the solenoid the greater will be the attraction and the higher the reading. (2) *Repulsion type*: two soft iron parallel armatures are magnetized by the same solenoid. Having similar polarities they tend to repel each other. One armature is fixed and one is attached to the pointer which is restrained by a spring. Both types of meter are equally suitable for d.c. or a.c. operation without the need for a meter rectifier. Unlike the moving-coil meter their scales are not uniform and they tend to lack sensitivity. They are more robust than moving-coil meters and for small currents no shunt resistor is required as the solenoid winding is sufficiently robust to sustain the circuit current.

muffle furnace (*mech*) A type of heat-treatment furnace where the heat is supplied to the outside of a refractory compartment (*muffle*) holding the work (*charge*). This prevents the products of combustion from contaminating the work. It also assists in ensuring uniform heating of the work. Separation of the combustion zone and the work also allows for *atmosphere control* within the muffle. In the electric resistance muffle furnace, where there are no products of combustion, the heating elements are themselves inside the muffle.

multiple (*maths*) A number that contains another number an exact number of times. For example, 12 is a multiple of 2 which it contains exactly 6 times. It is also a multiple of 3 which it contains exactly 4 times, or a multiple of 4 which it contains exactly 3 times, and 6 which it contains exactly twice.

multiplexing (*elec*) The use of one channel for multiple signals by time division or frequency division. See: *time division multiplexing*; *frequency division multiplexing*.

multiplication law of probability (*stats*) This law is recognized by the use of the word '*and*' connecting the probabilities. Thus if p_x is the probability of event X happening and p_y is the probability of event Y happening, the probability of events X *and* Y happening is $p_x \times p_y$. This law can be used for any number of events connected by the word '*and*'.

multiplier (*elec*) See: *voltage multiplier*.

multi-tooth cutters (*mech*) As the name implies these are cutting tools with more than one cutting edge, for example saw blades, drills and milling cutters. The use of multi-tooth cutters increases the rate of metal removal without increasing the load on any one individual cutting edge. Unfortunately, the greater the number of teeth; the smaller space for chip clearance, and the greater the likelihood of cutter breakage. See: *single-point cutting tools*.

multivibrator (*elec*) A relaxation oscillator with two active elements (transistors or thermionic valves) in which one is conducting when the other is not. An astable multivibrator switches continually between the two states. An example is shown in Fig. M.8 where the lamps would flash on and off alternately. The speed of flashing (time constant) will depend on the values chosen for the resistors and capacitors. Monostable and bistable multivibrator circuits have to be triggered by an external signal.

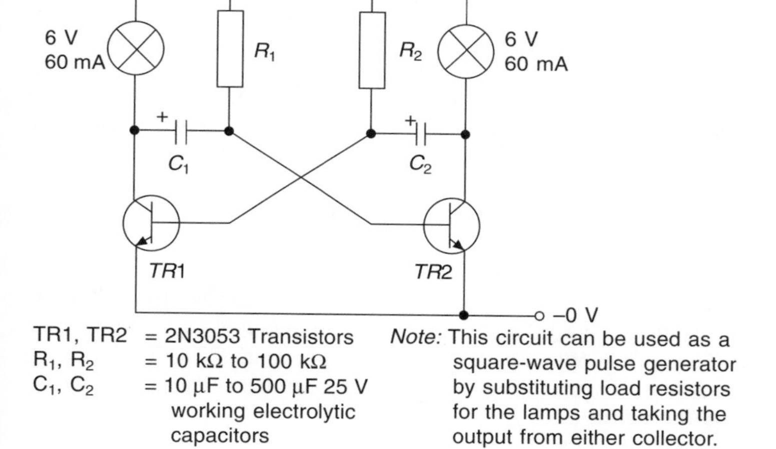

Fig. M.8 Astable Mutivibrator Circuit

muntz metal (or yellow metal) (*matls*) A brass alloy typically containing 60% copper, and 39% zinc with some lead and iron, used for products such as castings and hot-brass stampings for the plumbing industry. It is also used for hot extruded sections such as rods and tubes.

mutual inductance (*elec*) The induction of an e.m.f. in one circuit by a charging current in another; quantity symbol M, unit the henry (H). Two circuits have a mutual inductance of 1 henry if the e.m.f. induced in one circuit is 1 volt when the current in the other is changing at a rate of 1 ampere/second.

$$M = \frac{\text{induced e.m.f. in coil 2}}{\text{rate of change of current in coil 1}}$$

NAND gate (*elec*) A 'not-and' gate that combines the characteristics of an AND gate followed by a NOT gate. Figure N.1 shows its symbol and its truth table (compare this with the AND gate truth table).

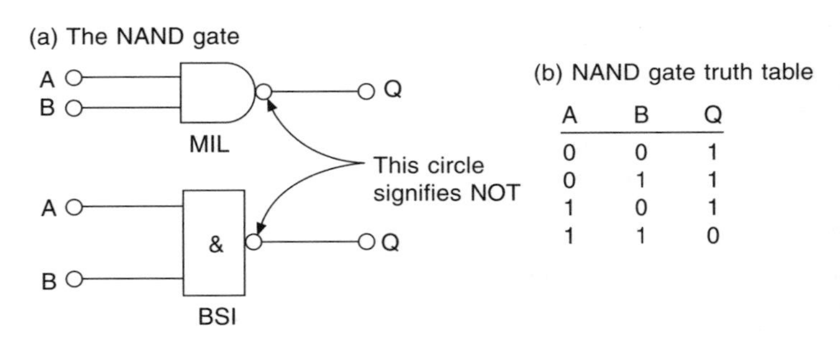

A	B	Q
0	0	1
0	1	1
1	0	1
1	1	0

Fig. N.1 NAND Gate and Truth Table

nanotechnology (*sci*) Study of the engineering of matter at a level close to that of individual molecules. Future predictions include swarms of *smart* nanorobots for medical treatment in the human body.

natural aspiration (*sci*) The method by which an internal combustion engine induces a fresh charge of air or air/fuel mixture into the combustion space by lowering the pressure in the combustion space below atmospheric pressure. Thus the air/fuel mixture is sucked into the combustion space, as opposed to *supercharged* engines where the air/fuel mixture is blown into the combustion space at a pressure above that of the atmosphere. The downward motion of the piston of a reciprocating engine on the induction stroke draws in a fresh charge of air/fuel mixture through the inlet valve.

natural frequency (*sci*) The frequency of vibration of a freely oscillating system, i.e. a system whose frequency of vibration is not subjected to external influences.

natural logarithm (*maths*) See: *logarithm*.

naval brass (*matls*) A brass alloy containing 62% copper, 37% zinc and 1% tin, so called because the tin content increases the alloy's corrosion resistance to sea water and reduces the chance of *dezincification* occurring. It is a corrosion-resistant form of *muntz metal* and is used where similar products are required for marine purposes. See: *Admiralty brass*; *dezincification*.

needle roller bearing (*mech*) Unlike a conventional roller bearing in which the rollers are located in a cage and roll between inner and outer races, in a needle roller bearing there is only an outer race which is attached to the bearing housing. The small-diameter (needle) rollers run directly on the shaft journal, which is hardened to prevent wear. An example is shown in Fig. N.2.

nested loop (*CAE*) This is a repetitive section of program contained (nested) within a subroutine. This is often referred to as a loop within a loop. For example, the subroutine may identify the positions of a cluster of holes in a repeated pattern. At each hole the nested loop will control the centre drilling, drilling and tapping of the holes with the accompanying tool changes. At the end of this sequence of operations the program moves back to the subroutine for positioning the next hole in the cluster. Thus the nested loop removes the need to repeatedly program the drilling and tapping routine for each hole position. See: *subroutine*.

neutral axis (*sci*) An imaginary longitudinal axis through a beam where there is no change of length and zero direct stress

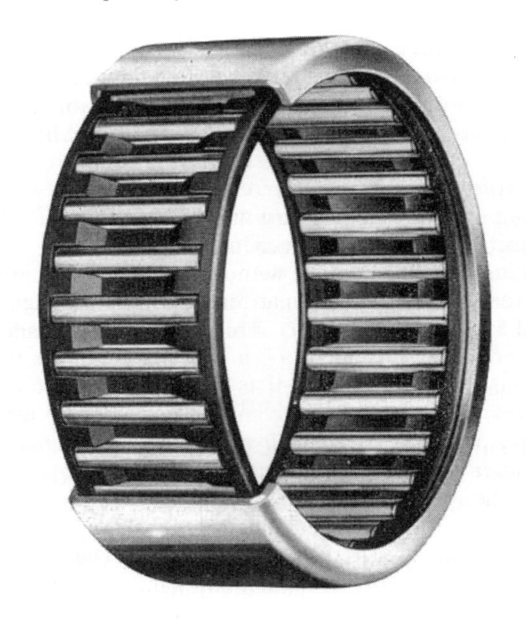

Fig. N.2 Needle Roller Bearing

when the beam is subject to bending. The material on one side of the neutral axis is subject to compressive stress and the material on the other side of the neutral axis is subject to tensile stress. For a rectangular section or any section symmetrical about a horizontal axis, this is theoretically at the mid-depth. In practice it is slightly offset owing to the fact that, for most materials, the tensile and compressive properties are not identical.

Newton–Raphson formula (*maths*) Also called Newton's formula, See: *iteration.*

Newton's laws of motion (*sci*) (1) A body continues in its state of rest, or uniform motion in a straight line, unless it is acted upon by an external force. (2) The rate of change of momentum of a body is proportional to the external force acting on the body, the change of momentum being in the direction in which the force acts. (3) For every force there is always an equal (reaction) force acting in the opposite direction.

nibble (*comp*) Four-bit binary number.

noise margin (*comp*) The maximum noise voltage that can be tolerated at the input of a gate without the output changing, where noise refers to unwanted or interference signal voltages.

nominal breaking stress (*sci*) This is defined in tensile tests by:

$$\text{nominal breaking stress} = \frac{\text{load at fracture}}{\text{original cross-sectional area}}$$

non-deterministic (*elec*) See: *stochastic*.

non-ferrous metals (*matls*) Metals and alloys other than those based on iron, e.g. copper, aluminium, brass, bronze, etc. See: *ferrous metal*.

non-flow energy equation (*sci*) $Q = W + \Delta U$ where Q is the heat transfer to the system, W is the work transfer from the system and ΔU is the change in internal energy, relating to closed thermodynamic systems where only energy crosses the boundary and not mass.

NOR gate (*elec*) A 'not-or' gate that combines the characteristics of an OR gate and a NOT gate. Figure N.3 shows its symbol and its truth table (compare this with the OR gate truth table).

normal (*sci*) At 90° to a surface.

normal distribution curve (*stats*) A symmetrical curve that can be described by a mathematical equation and that is of the shape shown in Fig. N.5. Many natural occurrences approximate to normal distribution. Tables of partial areas under the standardized normal curve are available to avoid repetitious calculations.

normalizing (*matls*) A heat-treatment process that is used to

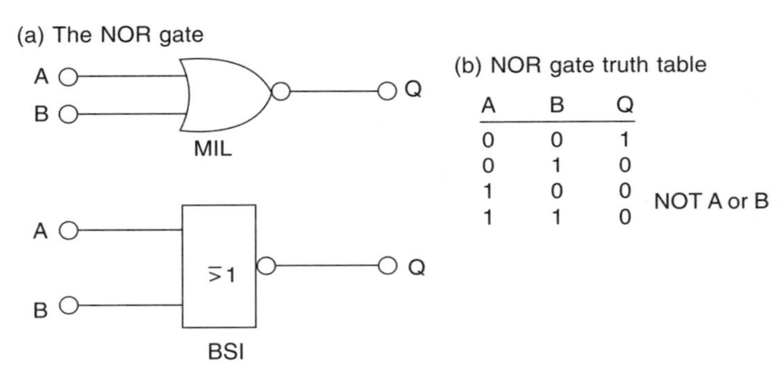

A	B	Q
0	0	1
0	1	0
1	0	0
1	1	0

Fig. N.3 NOR Gate & Truth Table

stress-relieve castings and forgings after rough machining so that they will remain dimensionally stable after finish machining. It involves heating a steel product to above its upper critical temperature until the material has a uniform temperature, then leaving it to cool to room temperature in still air in order to produce a uniform stress-free structure. Because of the more rapid cooling, the grain size of the metal after normalizing is more refined than after annealing and this results in the metal being tougher but less ductile. It is in an excellent condition for machining to a good surface finish.

Norton equivalent circuit (*elec*) A series or parallel network of resistors can be represented by a source that is a constant-current generator feeding an equivalent resistor R_{eq}, as shown in Fig. N.4. See: *Thevenin equivalent circuit*

NOT gate (*elec*) A single-input gate whose output is always the reverse of its input. It is an *inverter*. Fig. N.6 shows its symbol and its truth table.

nozzle (*sci*) A tube or corridor of varying cross-sectional area that guides the flow of a fluid to achieve an increase in fluid velocity and a corresponding drop in pressure.

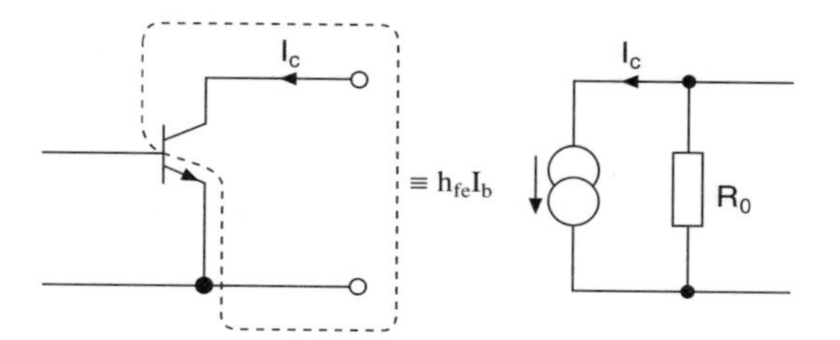

(a) Norton equivalent of the transistor output

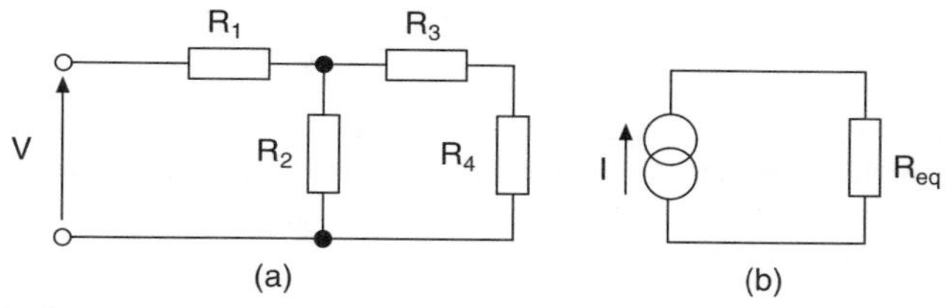

(b) Resistor network (c) Norton equivalent circuit

Fig. N.4 Norton Equivalent Circuits

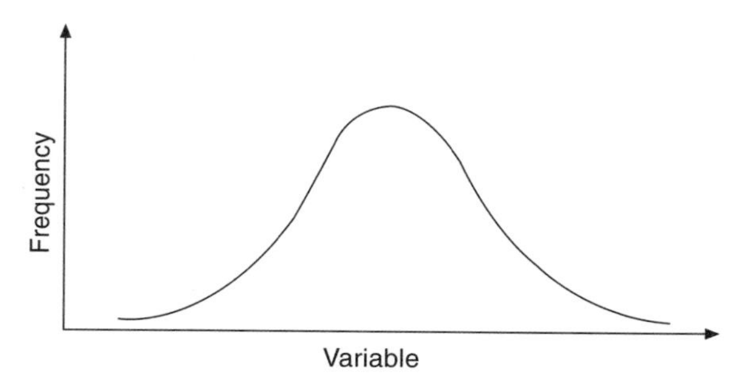

Fig. N.5 Normal Curve of Distribution

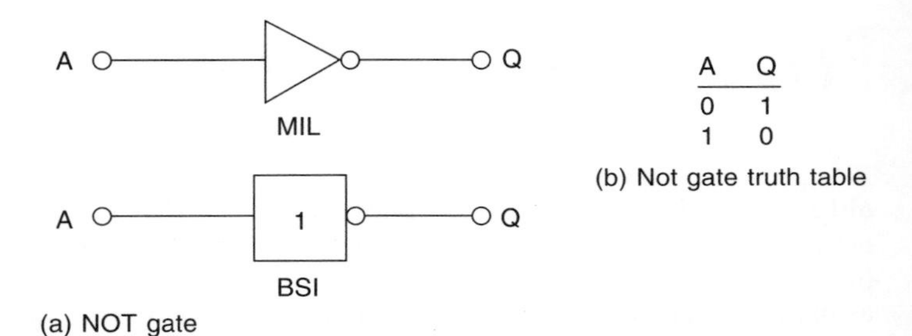

Fig. N.6 NOT Gate & Truth Table

numerator (*maths*) See: *fraction*.
Nyquist frequency (*maths*) A frequency equal to half the sampled frequency, for a sampled data system or a discrete time system. See also: *aliasing*.

O

obtuse-angled triangle (*maths*) See: *triangle.*
octal (*maths*) Base-8 numbering system. Each digit has a range of 0–7 and moving to the left each digit is worth 8 times as much as the digit immediately to its right. A single octal digit has the same range as a 3-bit binary number, enabling easy conversion between the two bases and representation of binary numbers. For example, to convert 613_8 to binary: $6_8 = 110$, $1_8 = 001$, $3_8 = 011$ giving a binary number of 110001011. See also: *hexadecimal.*
'off-line' programming (robots) (*CAE*) Has the same benefits as CNC simulation, i.e.

- The simulation is checked to ensure that the robot does not collide with any object within its cell.
- Expensive robots are not tied-up while the operator develops the robot program.
- Only one generic language is required for a number of different robots; the program can be post-processed into the required specific robot language by the computer software automatically.

ogive curve (*stats*) An S-shaped curve that is similar in appearance to a classical Gothic arch of the same name. See: *cumulative frequency distribution curve.*
ohm (*elec*) SI unit of electrical resistance, symbol Ω. If a conductor with a resistance of 1 Ω has a e.m.f. of 1 V applied to it, a current of 1 A will flow through the conductor.
ohmmeter (*elec*) Instrument for measuring electrical resistance. This instrument applies a known e.m.f. across a resistor and measures the resulting current flow through the resistor, assuming $I \propto 1/R$. An ohmmeter is usually included in a digital multimeter.

Ohm's law (*elec*) The current flowing through a metallic conductor at constant temperature in zero magnetic field is directly proportional to the potential difference across the ends of the conductor, i.e. $V \propto I$. The constant of proportionality in d.c. circuits is the resistance R, so that $R = V/I$.
open hearth furnace (*matls*) Obsolescent steel-making process where the charge is in contact with the products of combustion but out of contact with the fuel. The charge consists of pig-iron and steel scrap which is heated in an open shallow hearth.
open loop (*CAE*) Simple control system driven from the desired value only, and having no feedback from the actual output.
operational amplifier (*elec*) Basic electronic amplifier having a wide range of applications, with two inputs (one inverting and one non-inverting), one output, and a very high gain; used in feedback circuits.
optical fibres (*matls*) Fibres made from glass having high purity, low dispersion and low loss. They are used for the transmission of digital signals using pulsed light in telecommunications and in networking computers. In both applications one of their main advantages is their lack of susceptibility to electromagnetic interference and lack of corruption of the signal over long distances.
OR gate (*elec*) A logic gate which produces an output of logic 1 when either input is at logic 1. See: *truth table*; *logic gate*.
orifice flowmeter (*sci*) An instrument for measuring liquid flow in terms of the pressure drop across a calibrated constriction in a pipe.
origin (*maths*) The point where the axes of a graph cross, with coordinates (0, 0). Distances to the right of and above the origin are *positive* values, distances to the left of and below the origin are *negative* values.
oscillator (*elec*) Electronic device that produces an alternating current of a particular frequency and waveform. Ocillators are used in many electronic circuits, particularly cyclical measuring

instruments such as the oscilloscope, all computer equipment and digital instruments. They can be in a simple form designed to produce a regular pulse or at the other extreme, when used as a frequency standard, be capable of producing a particularly accurate and stable waveform.

oscilloscope (*elec*) An instrument basically consisting of a cathode ray tube, time-base generator and power pack used for visually displaying the waveform of electrical signals and allowing measurements of time, amplitude and frequency to be made. A visual image of the electronic signal is created by controlling a beam of electrons which strike a fluorescent screen. Horizontal deflection of the electron beam is provided by the internal time-base circuit. This causes the beam to sweep over the screen at a specific rate. The electronic signal is fed to the vertical beam control to produce a graph of signal versus time.

Otto cycle (*sci*) See: *four-stroke cycle*.

overcurrent protection (*elec*) System for the protection of an electrical circuit when an equipment fault causes the current flowing in the circuit to rise to unacceptable and dangerous levels. Overcurrent protection operates by disconnecting the fault from the circuit by means of a *fuse* or an *overcurrent relay*.

overcurrent relay (*elec*) Relay used in electrical protection systems. Two types are common: (1) Instantaneous, which operates when the current exceeds the set value of the relay; (2) inverse definite minimum time (IDMT), which has a minimum time during which an overload current must flow before the relay trips to ensure that it is not tripped unnecessarily by surges from, for example, motors starting up.

P

Paint (*matls*) Painting is a simple but effective way of preventing the corrosion of metallic components. A paint system consists of (1) a primer that adheres strongly to the metal surface to be coated, contains a corrosion inhibitor and provides a key for the subsequent coats to adhere to; (2) a second (middle) coat that builds up the colour and gives it 'body'; (3) a topcoat that contains a varnish to seal the system and prevent the absorption of water which could cause corrosion.
Pappus (*maths*) The *first theorem* of Pappus states that if a plane curve is rotated about an axis in its own plane so that it does not intersect that axis, then the surface area generated is given by the product of the curve length and the distance moved by the centroid of the curve during a single rotation. The *second theorem* of Pappus states that if a plane area is rotated about an axis in its own plane so that it does not intersect that axis, then the volume of the solid generated is given by the product of the area and the distance moved by the centroid of the area during a single rotation.
parabola (*maths*) The locus of a point that moves so that the distance from a fixed point (called the focus) is equal to the perpendicular distance from a line (called the directrix). For a parabola that is symmetrical about the x-axis, $y^2 = 4ax$ where a is the distance from the origin to the focus.
parallax error (*sci*) An instrumentation error due to reading the needle of an analogue display at an angle and lining it up with an incorrect graduation mark on the scale. A mirror is often placed behind the needle in the plane of the scale so that the observer can ensure that the line of vision is perpendicular to the scale and immediately above the needle. This occurs when the needle appears to be immediately above its reflected image.

parallel axis theorem (*sci*) A technique used for finding the second moment of area or moment of inertia of a body about an axis (I_{XX}) parallel to and other than the axis passing through the centroid (I_G):
(1) The second moment of area about axis *XX* is given by $I_{XX} = I_G + Ah^2$, where *A* is the area and *h* is the perpendicular distance from the centroid to axis *xx*
(2) The moment of inertia about *XX* is given by $I_{XX} = I_G + Mh^2$, where *M* is the mass and *h* is the perpendicular distance between the axes.

parallel connection (*elec*) A method of connecting the elements of an electrical circuit so that the applied e.m.f. is common to all the limbs of the circuit, as shown in Fig. P.1. The total current flowing is the sum of the currents in the individual circuits. Should one limb of the circuit fail the remaining limbs will be unaffected. See: *series connection.*

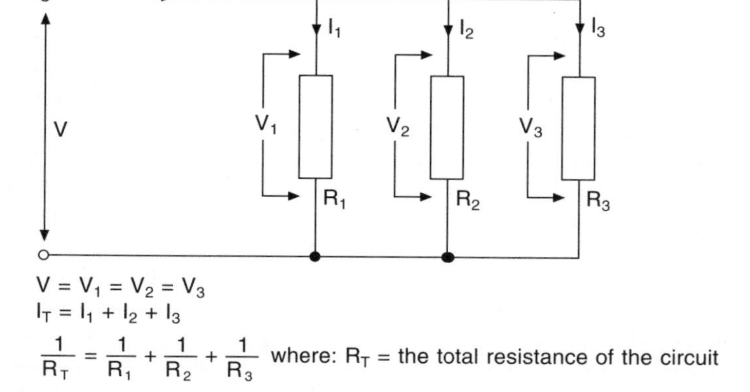

$V = V_1 = V_2 = V_3$

$I_T = I_1 + I_2 + I_3$

$\frac{1}{R_T} = \frac{1}{R_1} + \frac{1}{R_2} + \frac{1}{R_3}$ where: R_T = the total resistance of the circuit

Fig. P.1 Parallel Connection

parallel lines (*maths*) These are lines that lie in the same plane and are equidistant from each other so that they do not meet.
parallelogram (*maths*) A four-sided figure in which the opposite angles are equal but not right angles, opposite sides are equal in length and parallel, and the diagonals bisect each other but are not equal in length.
parallel path system (*CAE*) A system in which the tool moves from one position to the next so that the tool path is always in a straight line parallel to the x– or y-axis. The traverse rate is under the control of the programmer and this system can be used for simple milling operations but is unsuitable for contouring and profiling. Both point-to-point and parallel path systems are known as linear path systems. See: *continuous path* and *point-to-point system*.
parallels (*mech*) A pair of identical, rectangular, hardened-iron or steel supports, with adjacent sides square and opposite sides parallel, used for supporting work parallel to a datum surface when marking out or machining. They are available in matched pairs in a range of sizes.
parity (*CAE*) A means of distinguishing between punched tapes prepared to EIA standards and punched tapes prepared to ISO standards. See: *punched tapes*.
Parkerizing (*matls*) See: *conversion coatings*.
partial fractions (*maths*) A technique whereby a complex fraction is expressed as the sum of a number of simpler fractions. These are called *partial fractions* because each is part of a whole. For example, $2x/(x^2 - 1)$ can be written as $[1/(x + 1)] + [1/(x - 1)]$. The basic rules are: (i) the denominator must factorize; (ii) the numerator must be at least one degree of order less than the denominator.
parting-off (*mech*) Operation carried out on a lathe, after the work has been completed, to cut the finished workpiece from the rest of the bar of material. This is carried out using a narrow cutting tool called a *parting-off tool* held in the toolpost of the

machine and fed radially into the workpiece. See: *single-point tools*.

pascal (*sci*) SI unit of pressure, symbol Pa; 1 Pa = 1 N/m^2.

Pascal's laws (*sci*) Three laws relating to fluid pressure were promulgated by Blaise Pascal in the seventeenth century. (1) For a fluid at rest, the pressure is the same throughout if the weight of the fluid is neglected. (2) The static pressure acts equally in all directions at the same time. (3) The pressure always acts at right angles to any surface in contact with the fluid.

Pascal's triangle (*maths*) A convenient summary of the coefficients in the binomial expansion of $(a + b)^n$, where n is an integer as shown in Fig. P.2. Each row relates to a particular value of n, and each coefficient is determined by adding together the two coefficients immediately above it on the previous line.

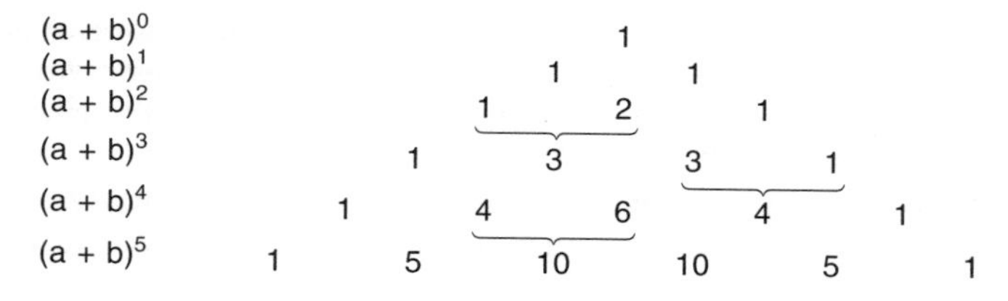

Fig. P.2 Pascal's Triangle

pattern (*mech*) Facsimile of an object made of wood, metal or resin-bonded fibreglass for making the *mould* in *foundry* work. The pattern must be made slightly larger that the final casting to allow for shrinkage as the metal cools. The pattern shape must taper where appropriate to allow removal from the finished mould without disturbing the sand. See: *pattern-maker*; *contraction rule*.

pattern maker (*mech*) A highly skilled craftsperson who makes *casting patterns* for use in a *foundry*. The pattern maker must also decide on the orientation of the pattern to allow the molten metal to flow correctly, and where the joints should be in the mould.

PC (*comp*) See: *personal computer*.

pearlite (*matls*) A constituent of steels and cast irons in which the grains have a lamellar structure of alternate layers of ferrite and cementite (iron carbide) formed simultaneously from austenite upon cooling. This is the toughest structure in steels.

pendulum (*sci*) A simple pendulum consists of a small dense mass suspended from a fixed fulcrum by a thread of negligible weight. The frequency, f, of oscillation when swinging freely with small amplitude is given by:

$$f = \frac{1}{2\pi}\sqrt{\frac{g}{l}}$$

where g is the acceleration due to gravity and l is the length of the thread. Notice that the frequency of oscillation for any position on the Earth depends on the length of the thread and is independent of the mass.

PEPS (*CAE*) PEPS (production engineering productivity system) is a powerful computer-aided manufacturing (CAM) system with one-line commands defining pockets and profiles. It can import DXF files which, in turn, can be defined as part boundaries (outer profiles and inner pockets slots) called K curves. Once the program has been proved (the simulation cutter paths are correct), then the simulation can be post-processed into any CNC language such as Heidenhain, Fanuc, Allen Bradley, Phillips, etc.

percentage (*maths*) These are fractions having the number 100 as their denominator in order to give them a common standard. Thus 1/2 = 50/100 = 50%.

percentage elongation (*sci*) As derived from a tensile test on a material it can be calculated using the expression:

percentage elongation

$$= \frac{\text{length at fracture} - \text{original length}}{\text{original length}} \times 100$$

$$= \frac{\text{elongation} \times 100}{\text{original length}}$$

percentage reduction in area (*sci*) Derived from a tensile test it can be calculated using the expression:

percentage reduction in area

$$= \frac{\text{original CSA} - \text{CSA at fracture}}{\text{original CSA}} \times 100$$

where CSA is the cross-sectional area.

percentage relative frequency (*stats*) See: *relative frequency.*

percentiles (*stats*) See: *quantiles.*

perfect flow (*sci*) Ideal flow regime of fluid through a pipe whereby every particle is assumed to flow at the same velocity and in a straight line parallel to the walls of the pipe. Although not existing in reality, it is useful for calculation purposes.

perfect gas (*sci*) An ideal, hypothetical gas that obeys the ideal gas law $pV = nRT$ under all conditions, where p = pressure, V = volume, n = the number of moles, R = the gas constant per mole and T = the absolute temperature. It is comprised of molecules of negligible volume that exert no forces between themselves, collisions between the molecules and the container walls being perfectly elastic. In reality the behaviour of gases tends towards that of a perfect gas at high degrees of superheat and low pressures. See: *gas laws.*

perfectly elastic collision (*sci*) A hypothetical collision between two elastic bodies with no energy loss due to heat, sound, etc. where the total kinetic energy before the collision is equal to the

total kinetic energy after the collision. See also: *principle of conservation of momentum*.
periodic functions (*maths*) A function f(x) is said to be periodic if f($x + T$) = f(x) for all values of x when the time interval T is a positive number. In this context, T is the interval between two successive repetitions and is called the period of the function f(x). For example $y = \sin x$ is periodic with a period of 2π (one complete cycle).
periodic signal (*sci*) Signal with a basic form repeated continuously. See also: *periodic time*.
periodic time (*sci*) The time for one complete oscillation of any dynamic action that is continuously and identically repeated.
Peristaltic pump (*sci*) A pump whose action is caused by rollers moving in succession over a flexible tube as shown in Fig. P.3; often used in medical situations as the liquid in the tube can remain isolated and sterile.

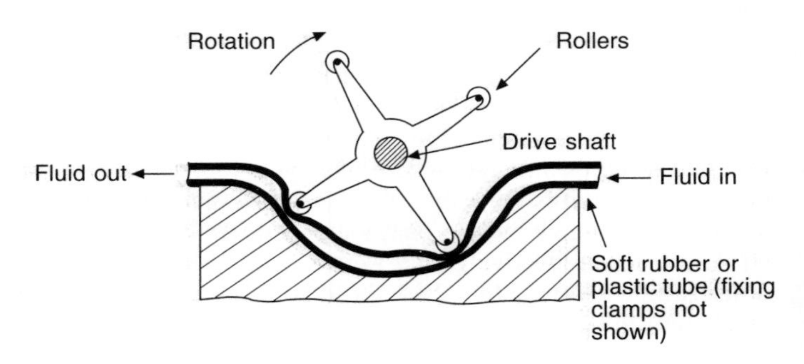

Fig. P.3 Peristaltic Pump Principle

permeability (*elec*) The ratio of the magnetic flux, B, in a magnetic material to the magnetic field strength, H, quantity symbol

μ, unit henry/metre (H/m); $\mu = B/H$. If the magnetic field exists in a vacuum then the value of μ is a constant represented by μ_0 and equal to $4\pi \times 10^{-7}$ H/m. The relative permeability μ_r is the ratio of the magnetic flux density produced in a material to the value for free space. Absolute permeability can then be calculated from $\mu = \mu_0\mu_r$.

permittivity (*elec*) The ratio of the electric flux produced, D, to the electric field strength, E; quantity symbol ε, unit farad/metre (F/m); $\varepsilon = D/E$, a property associated with dielectric materials used in capacitors. If the electric field exists in a vacuum then the value of ε is a constant represented by ε_0, which is called the permittivity of free space, and is equal to 8.854×10^{-12} F/m. For a dielectric, the relative permeability $\varepsilon_r = \varepsilon/\varepsilon_0 = C/C_0$, the ratio of the capacitances of a capacitor with and without that dielectric.

perpendicular (*maths*) At right angles to a plane surface or line.

personal computer (*comp*) An independent *computer* (although it could be part of a network) with *ROM*, *RAM*, *power pack*, a *central processing unit*, *keyboard*, and *VDU*.

phase (*elec*) That fraction of a periodic waveform, e.g. a sinusoidal waveform, that has been completed at a specified reference time. It is expressed as an angle where one cycle represents 360° (2π radians). The terms *in phase*, *in quadrature* and *in anti-phase* relate to two signals whose phase angles differ by 0° (or 360°), 90° and 180°, respectively. See: *state*.

phase angle (*elec*) (or phase difference) A measure of how much one sine wave leads or lags behind another reference wave of the same frequency, usually refering to a.c. electric circuits. The general standard expression for an a.c. voltage is $V = V_m \sin(\omega t \pm \phi)$, where V_m is the amplitude, ω is the angular frequency, t is the time and ϕ is the relevent phase angle. Thus two voltage signals, V_1 and V_2 with a phase difference of ϕ radians could be described by:

$$V_1 = V_{m1} \sin\omega t$$

$$V_2 = V_{m2} \sin(\omega t \pm \phi)$$

phase modulation (*elec*) *Modulation* where the phase of the periodic carrier wave is varied to represent changes in the amplitude of the modulating signal.
phasor diagram (*elec*) Diagram in which alternating currents and voltages are represented by vectors considered to rotate about an origin at a constant velocity anticlockwise. The length of the phasor represents the r.m.s. value and the angle with respect to the reference axis represents the phase angle. Voltage is usually the reference quantity and is positioned at 0° (i.e. the 3 o'clock position). Any a.c. quantities can be represented provided that all the waveforms are sine waves and all have the same frequency on the same diagram. This type of diagram is a much simpler representation of a.c. quantities than mathematical formulae.
photoconductive cell (*elec*) A sensor (photodiode) in which the electric current flowing though the component is proportional to the light intensity falling on it so that the electrical resistance reduces as the light intensity increases. This type of cell is useful in instrumentation when direct contact with a process is difficult but changes in the light intensity can be related to the quantity to be measured.
pictograms (*stats*) See: *ideographs.*
pie diagram (*stats*) A graphical representation in which the whole is represented by a circle and the areas of the sectors into which it is divided are proportional to the data that makes up the whole as shown in Fig. P.4.
piezoelectric effect (*elec*) The generation of a potential difference across the faces of particular types of crystal when subject to mechanical strain; alternatively, the crystal experiences mechanical deformation when an electric field is applied to it. The electric field is directly proportional to the stress, and if the polarity of the

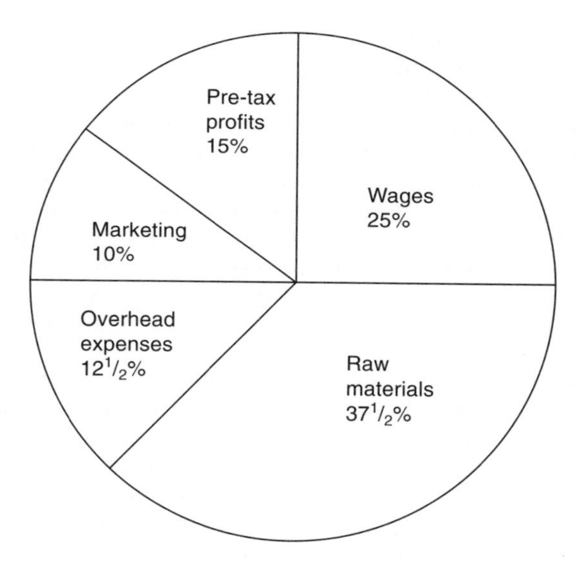

Fig. P.4 Pie Diagram

applied field changes, the stress changes from compression to tension.

pig-iron (*matls*) Impure iron from a blast furnace that is cast into pigs after smelting iron ore to extract metallic iron. Originally the molten iron from the blast furnace was run directly into moulds formed in the sand floor of the iron works. The main channel was called the 'sow' and the side moulds feeding from the sow were called the 'pigs' (piglets). Refined pig-iron is the basic material used in an iron foundry. At one time cast-iron 'pigs' were used for steel manufacture but in a modern, integrated steel works, the molten iron from the blast furnaces is used directly for steel making without cooling and solidifying. The molten product of the blast furnace is still referred to as pig-iron.

pipe thread (*mech*) See: *British Standard Pipe thread.*

piston (*mech*) **1** A sliding component within the cylinder of a reciprocating engine having a pressure-tight seal between the piston and cylinder walls. The piston converts pressure energy from the working fluid into mechanical energy. This is achieved by the pressure of the working fluid imparting linear motion to the piston that, in turn, produces a force on a connecting rod that rotates the crank shaft. **2** A similar sliding component in a compressor that converts mechanical energy into pressure energy in the working fluid.

piston ring (*mech*) Circular metal split rings housed in circumferential slots in the piston of a reciprocating engine or compressor and which create a gas-tight seal between the piston and cylinder walls by flexing radially outwards. The rings also control the lubrication of the cylinder walls by regulating the oil that passes into the combustion space and by distributing a film from oil that is splashed into the cylinder walls.

pitch (*mech*) **1** The linear distance between identical points on adjacent threads of a screw measured parallel to the axis. (See: *lead* (*mech*)) **2** The linear distance between the centres of adjacent hole positioned on a pitch circle.

pitch circle (*mech*) **1** The pitch circles of a pair of intermeshing gears are imaginary circles around which the pitch of the teeth are measured and which form the basis for the calculations of the tooth proportions. When running on the same centre line as the gears the pitch circles would revolve together at the same speed without slipping. **2** A circular centre-line datum on which the bolt-holes of a flange are positioned.

planimeter (*stats*) An instrument for measuring small areas bounded by irregular curves and straight lines.

planned maintenance (*mech*) The planned or *preventative* *maintenance* of plant and equipment, carried out according to a plan based on time intervals or monitoring in order to minimize *breakdown maintenance* which is generally more expensive.

plant (*mech*) A general term referring to the machinery, equipment or sytems by which processes are carried out. Plant can be mechanical, thermal, electrical, construction, etc.

plasticity (*sci*) Property of materials that describes the change in shape (plastic deformation) that can occur as the result of an applied stress greater than the stress at the elastic limit. In theory the material retains its new shape after the applied force is removed. In practice there is always some elastic component present that causes a slight 'spring-back' as the material tries to revert to its original state. See: *elasticity*.

plumb-bob (*mech*) Simple but effective instrument comprising a mass with a downward point suspended on a line, used for vertical alignment with the top of the line to the point of the bob. Theoretically the plumb-bob points to the centre of the Earth and therefore all lines described by the plumb-bob are radial and not parallel. However, because of the large diameter of the Earth compared with the distance between adjacent lines set with a plumb-bob, such lines can be considered as parallel for most practical purposes.

pneumatics (*sci*) Systems for the transmission and control of power using compressed air as the working fluid. See: *hydraulics*.

point-to-point system (*CAE*) A system that positions the tool at a specific point. The path the tool takes between the points and the traverse rate is neither under the control of the programmer nor the operator. The path taken is usually the shortest. Such a system is suitable only for simple drilling operations or sheet metal turret punching. See: *continuous path*; *parallel path systems*.

Poisson distribution (*stats*) The Poisson distribution is used in preference to the binomial distribution when a large number of trials, n, are involved (normally greater than 10). If n is large, p

is small and pn is less than 5, then a very good approximation to the binomial distribution can be given by the Poisson expression. The probabilities that an event will occur 0, 1, 2, 3, …, n times in n trials are given by successive terms of the Poisson expression:

$$e^{-\lambda}\left(1 + \lambda + \frac{\lambda^2}{2!} + \frac{\lambda^3}{3!} \cdots\right)$$

taken from left to right.

Poisson's ratio (*sci*) Ratio of lateral strain to longitudinal strain when a material is stretched, symbol ν. If the material width is w and the length is l, then lateral strain = $\Delta w/w$ and longitudinal strain = $\Delta l/l$. Then:

$$\text{Poisson's ratio, } \nu = -\frac{\Delta w/w}{\Delta l/l}$$

where Δw = increase in width and Δl = increase in length. For example, for steel $\nu = 0.29$.

polar second moment of area (*sci*) The second moment of area about a fixed point rather than an axis, symbol J, units m^4. For a shaft, the polar second moment of area represents a measure of the resistance to twisting.

For a solid shaft

$$J = \frac{\pi D^4}{32}$$

For a hollow shaft

$$J = \frac{\pi(D^4 - d^4)}{32}$$

where D is the outer diameter and d the inner diameter.

See also: *moment of area*; *torsion equation.*

polygon (*maths*) Any enclosed plane figure bounded by straight lines. A triangle has three sides, a quadrilateral four, a pentagon five, a hexagon six, etc.

polygon of forces (*sci*) Representation of *concurrent forces* as vectors drawn head to tail as shown in Fig. P.5. Each vector is in the same direction as the force it represents, with a length proportional to the magnitude of that force. If the diagram closes then the forces are in equilibrium; if the diagram is not closed

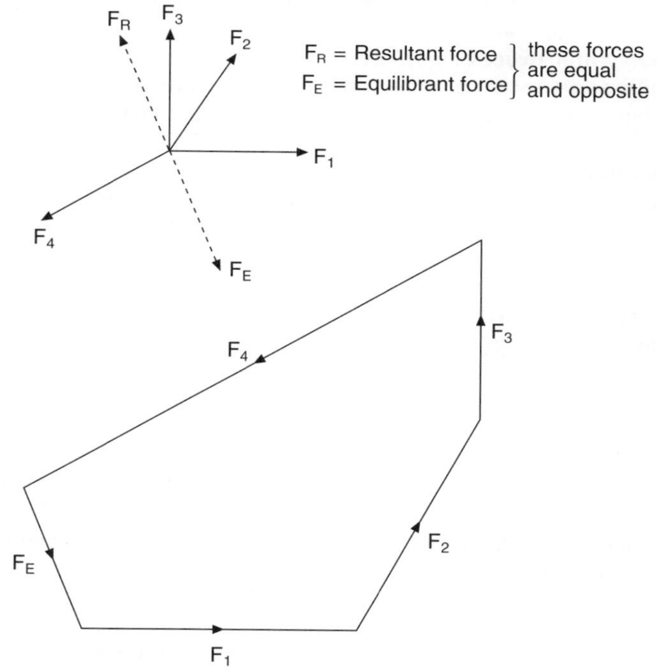

Note: the force F_E is needed to 'close' the polygon and bring it into equilibrium

Fig. P.5 Polygon of Forces

then the vector that is required to close it is proportional to and in the same direction as the equilibrant.

polytetrafluoroethelyne (PTFE) (*matls*) A thermoplastic used in engineering as it is chemically inert and has the lowest known coefficient of friction. An example of use is for anti-friction slideways on CNC machines and for non-lubricated bearings in some small compressors and office equipment.

polytropic process (*sci*) Reversible process in thermofluids that follows pV^n = a constant, where p = pressure and V = volume. Many non-flow processes of vapours and gases can be approximated to this law.

population (*stats*) A set containing all the possible members.

potential difference (*elec*) Difference in voltage levels between two points in a circuit owing to current flowing through a resistance, measured in volts (V); not to be confused with *e.m.f.* which causes the current to flow.

potential divider (*elec*) See: *voltage divider.*

potential energy (*sci*) Energy of a body or a system owing to its position, state or shape; this includes electrical, nuclear and chemical energy. In mechanics a body raised to a height h above the ground has potential energy equal to mgh where m is the mass of the body, and g is the acceleration due to gravity.

potentiometer (*elec*) **1** A precision measuring instrument in which an unknown *e.m.f.* or *potential differences* is balanced against an adjusted *potential* provided by a constant current derived from a standard cell. Balance is indicated when a centre zero *galvometer* in the circuit indicates a null reading. **2** A three-terminal potential divider in which the resistance ratio is varied by linear movement of a slider or rotation of a spindle. The change in resistance ratio may follow various laws, e.g. logarithmic, linear, cosine, etc.

power (*sci*) Rate of doing work or using energy, quantity symbol P; the SI unit is the watt (W), where 1 W = 1 J/s.

power factor (*elec*) Ratio of the true power of an electric circuit to the total volt-amperes (apparent power); the ratio is equal to cosine ϕ where ϕ is the phase angle.

powers (*maths*) When an index is an integer it is called a *power*. Thus 3^5 is called 3 to the power of 5. The exceptions are when the indices are 2 and 3. Thus 3^2 is called 3 squared, and 3^3 is called 3 cubed.

precision (*mech*) Property of a system when the process is in a steady state, which indicates how close repeated readings are to each other. An instrument of low precision has readings scattered over a wide range when measuring the same value of a constant quantity. An instrument of high precision measuring the same value of the same quantity will produce readings spread only over a small range. The ability to manufacture large quantities of the same components with minimal dimensional variation. Precision should not be confused with *accuracy*. See: *accuracy*.

pre-set tooling (*mech*) Single-point lathe toolholders that have adjustable setting pads. They are used where suitable qualified tooling is not available, appropriate or sufficiently accurate. Pre-setting is carried out in a setting fixture away from the machine in advance of requirement.

press (*mech*) A machine tool for providing a closing force on cutting and forming tools. Manually operated presses have a multistart screw thread or a toggle mechanism to magnify the effort. Power-operated presses are classified according to the method of applying the closing force. The main classifications are: crank press, friction screw press, toggle press, pneumatic press and hydraulic press. Presses are rated in tonnes or in tons, the closing force being equivalent to the force of gravity acting on the rated mass.

press forging (*mech*) A process similar in principle to *drop forging* except that a *forging press* is used instead of a drop hammer. Press forging also uses top and bottom dies but only one blow is used for each forging in each die impression. The forgings are usually made in two or three progressive stages. The equipment cost is higher for press forging than for drop forging but the production rates are higher and less skill is required. The process can also be automated.

pressure (*sci*) The force that is exerted over a specific area, quantity symbol p, SI unit the pascal (Pa), so that $p = F/A$, where F is the force being exerted and A is the area upon which the force is acting. 1 Pa = 1 N/m^2. Also the *bar* is a unit commonly used in meteorology for atmospheric pressure, where 1 bar = 10^5 N/m^2.

pressure gauge (*mech*) See: *bourdon tube*.

pressure head (*sci*) See: *head of pressure*.

primary quantities (*maths*) See: *base units*.

principle of conservation of energy (*sci*) See: *conservation of energy*.

principle of conservation of momentum (*sci*) See: *conservation of momentum*.

principle of moments (*sci*) For a body at rest subjected to a number of *forces*, there must be a balance of *moments*. For a state of equilibrium, the clockwise moments about any point must equal anticlockwise moments about that point. Expressed mathematically, the algebraic sum of the moments about any point must equal zero.

prints (*mech*) The pockets left in the mould by projections on a casting pattern for locating and supporting *cores* when casting hollow articles.

prism (*maths*) Any solid of constant, regular cross-section. Volume = length × cross-sectional area.

prismoidal rule (*maths*) See: *volume of solids*.

probability (*stats*) The likelihood of an event occurring. A concept which seeks to quantify the relative frequencies of events by answering the following question: if a given coincidence of circumstances takes place a large number of times, on what proportion of these occasions would a particular event occur? The outcome is expressed as a proper or decimal fraction. For example, a die has six sides numbered one to six, therefore the chances of throwing a four is 1/6 or 0.16667. Therefore the probability of an event happening (p) is 1/6 and of it not happening (q) is 5/6; $p + q$ always equals unity ($p + q = 1$)

process control (*mech*) **1** The computerized, automated control of process industries such as the petrochemical and food industries, where the plant dynamics are usually uncertain and fixed compensators are not normally used. General-purpose controllers that can be tuned to suit the plant dynamics are more popular, e.g. *PID controllers*, used in *feedback* loops where plant measured values are compared with the desired values to produce an error signal to drive the controller. For simplicity, quantities are usually measured as a percentage rather than specific units as a process can involve a large number of different variables. **2** The management of industrial processes.

product (*maths*) The numerical result of multiplying two or more quantities together.

program format (*CAE*) Different control systems use different formats for the assembly of each block of data. Always consult the programming manual for the machine being programmed. A block of data consists of a complete line on a program containing a complete set of instructions for the controller. There are two systems of program format: the obsolescent *fixed block* (*sequential*) *format* system, and the currently used *word* (or *letter*) *address* system. See: *word address system*; *fixed block (sequential) format system.*

programmable logic controller (**PLC**) (*CAE*) A digital electronic controller based on computer logic and used for controlling engineering devices and process systems. The controller can be programmed by the operator to carry out a specific sequence of interrelated operations, and reprogrammed at will to suit any changes to that requirement. A simple example is the control unit for traffic lights where they can be programmed to change the sequence and time intervals at which the lights operate to meet the requirements of changing traffic patterns.

propagation delay time (*elec*) The time taken for the output of a gate to change after an input has been applied.
proper fraction (*maths*) A fraction in which the numerator is smaller in value than the denominator. See: *fraction*.
proportion (*maths*) If one quantity is *directly proportional* to another quantity, then any change in one results in a corresponding change in the other; e.g. if a is proportional to b and b is trebled then a is also trebled. If one quantity is *indirectly proportional* to another quantity, then when one is doubled, the other is halved. The sign of proportionality is a colon (:). Thus if 100 g of a substance is shared out in the proportion of 1 : 4, one portion will weigh 20 g and the other will weigh 80 g.
pulley (*mech*) A wheel that forms part of a belt drive power transmission system. The pulley wheel may drive or be driven by the belt. Grooved pulleys are used in conjunction with V-belts, toothed pulleys are used in conjunction with toothed (non-slip) belts, and crowned pulley wheels are used in conjunction with flat belts. See: *belt drives*.
pulley block (*mech*) Simple machine consisting of wheels that can rotate freely in a block, designed to carry a rope or chain for lifting loads. A rope pulley system usually consists of two pulley blocks, one at the top suspended from a fixed point such as a beam and the other at the bottom attached to a load. Either there are the same number of wheels in each block or one block has one more wheel than the other. A rope is threaded over each pulley in turn. One end is fastened to the block opposite the last pulley, the other is for applying effort (the tail rope), usually the effort acts in the opposite direction to the load. The velocity ratio is equal to the number of ropes passing from one block to the other or, when the effort acts in the opposite direction to the load, to the number of wheels in the system.
punch (*mech*) Hand or machine tool for marking, shaping or cutting holes in metal. Different shapes and forms are used, e.g.

a centre punch marks the centre point for starting drilling, a rivet punch is used for forming a rivet head, etc.

punched tape (*CAE*) This was the original method of loading program data into numerically controlled machines. Early machines, before the introduction of computerized controllers were called tape-controlled machines. With the advent of computer numerical control (CNC) punched tapes were used to load previously programmed data into the machine's control unit to reduce idle-time. Two systems were and still are available, both using 25-mm wide tape and each character is represented by a pattern of holes punched across the width of the tape. In the Electrical Industries Association (EIA) system each row has an *odd number of holes*. If a character only requires an even number of holes, then an extra hole has to be punched in track 8. This track is called the *parity track*. In the ISO system each row has to have an *even number of holes*. If a character only requires an odd number of characters then an extra hole has to be added in parity track 8. Thus the EIA system is referred to as an *odd-parity* system and the ISO system is referred to as an *even-parity* system. Examples of tape punched in both systems are shown in Fig. P.6 See: *conversational data input*; *manual data input*; *magnetic tape input*; *direct numerical control*.

pyrometer (*sci*) An instrument for measuring high temperatures that are beyond the range of a mercury-in-glass thermometer. The most common pyrometer uses a thermocouple as the sensing element to produce an electric current that is proportional to the temperature. A sheathed thermocouple may be inserted directly into the furnace or the radiant heat from the furnace may be focussed onto the thermocouple by means of a parabolic mirror (radiation pyrometer). See: *Thermocouple*.

Pythagoras's theorem (*maths*) For a right-angled triangle, the square of the length of the hypotenuse (side opposite the right angle) is equal to the sum of the squares of the lengths of the other two sides. For example, this can be shown to be true for a triangle whose side lengths are in the ratio 3 : 4 : 5 since $3^2 + 4^2 = 25 = 5^2$.

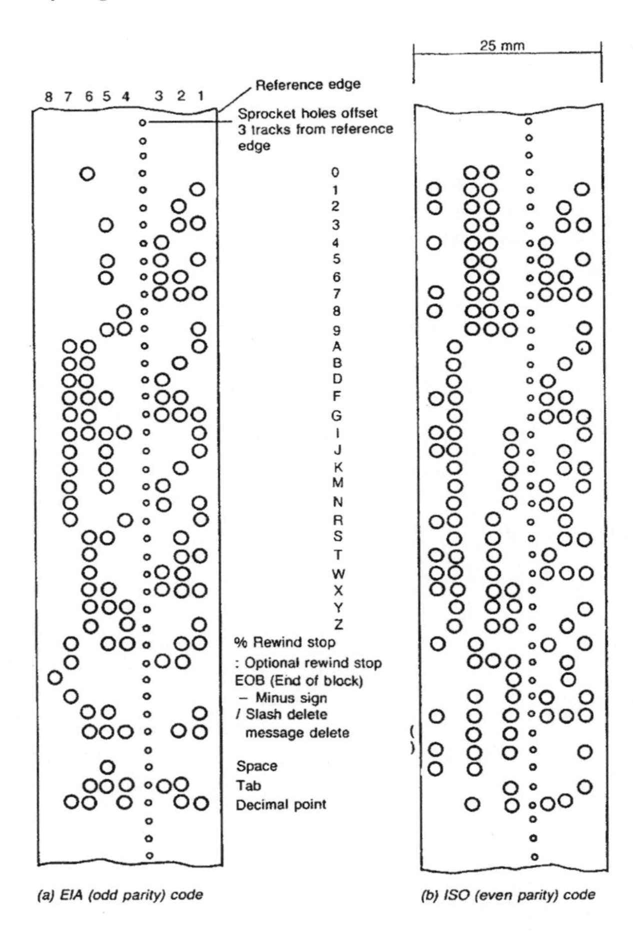

Fig. P.6 Even and Odd Parity Tapes [MTI, Fig. P.6]

quadratic equation (*maths*) An algebraic equation of the second order, where the variable x is raised to the power of 2, giving an equation of the form $y = ax^2 + bx + c$, where a, b and c are constants; its solution is given by

$$x = \frac{-b \pm \sqrt{b^2 - 4ac}}{2a}$$

The graph of this equation is a parabola. Higher-order equations are possible and follow a similar pattern. Thus a third order equation would be written $y = ax^3 + bx^2 + cx + d$ and so on.

quadrilateral (*maths*) Any four-sided figure. The internal angles of any quadrilateral add up to 360°. See: *rectangle*; *square*; *parallelogram*; *rhombus*; *trapezium*.

qualified tooling (*mech*) Single-point lathe tools made to ISO standard 1832 so that the distances from the datum surfaces of the tool shank to the cutting edges of the tool tip are maintained to within ± 0.08 mm. This enables tools to be replaced with a minimum of adjustment to the machine. Such tools have disposable cutting tips that are also made in bulk at low cost to high levels of accuracy and repeatability. It is quicker and cheaper to replace the tip than to regrind a conventional brazed-tip tool.

Quantiles (*stats*) The three values that divide a set of discrete data into four equal parts or subsets are called *quartiles*. For large sets the nine values that divide the data into ten subsets are called *deciles*. Very large sets may be divided into one hundred subsets, where the equivalent values are known as *percentiles*.

quantity symbol (*maths*) Algebraic symbol for representing a quantity and for mathematical manipulation. The SI system uses

standard quantity symbols, e.g. mass, m, volume, V. Quantity symbols are usually shown in an italic font to differentiate them from unit symbols.
quartiles (*stats*) See: *quantiles*.
quenching (*matls*) Process of cooling metal quickly as part of a heat-treatment process (hardening process for steel). Water is the usual cooling medium for plain carbon steels but quenching oil may be used to achieve a less brittle (tougher) result with less chance of cracking and distortion. Alloy steels have a lower critical cooling rate and are usually quenched in oil to achieve their maximum hardness. Lubricating oils must NOT be used for quenching because of the fire risk and the noxious fumes that will be given off. Only specially compounded quenching oils should be used. The quenching bath should be fitted with an airtight lid in case of fire.
quotient (*maths*) See: *division*.

radian (*maths*) Measure of an angle formed at the centre of a circle by a sector. When the arc of a sector is the same length as the radius, the angle at the centre of the circle made by the sector is equal to 1 radian. Therefore as the length of the circumference of a circle is equal to $2\pi r$, then in 360° there are 2π radians and 1 radian = $360°/2\pi = 57.296°$. In calculations radians are usually shown by a 'c' superscript although the radian is dimensionless, e.g. 1.3 radians → 1.3^c. The radian is very useful in engineering calculations of angular motion as the linear distance moved by rotating a body about an axis is equal to the product of the radius and the angle turned through in radians.

radius of gyration (*sci*) For a rotating body, the *radius of gyration*, quantity symbol k, is the distance from the axis of rotation of an imaginary point at which all the mass (m) of the body can be assumed concentrated. The moment of inertia of the body (I) can be calculated from the expression $I = mk^2$.

rake (*mech*) This is the tool angle that controls the chip formation and cutting efficiency of any cutting tool. It can have a positive, zero or negative angle as shown in Fig. R.1. The rake angle selected will depend upon the application and the cutting tool material. See: *metal-cutting tool angles.*

RAM (*comp*) Random access memory. A volatile computer memory that is lost when a computer is turned off. It is used to hold data or programs that may be loaded from storage. See also: *ROM.*

Rankine cycle (*sci*) An ideal cycle upon which the efficiency of steam plants are based. It comprises the introduction of pressurized water to the boiler, the evaporation of the water to steam, adiabatic

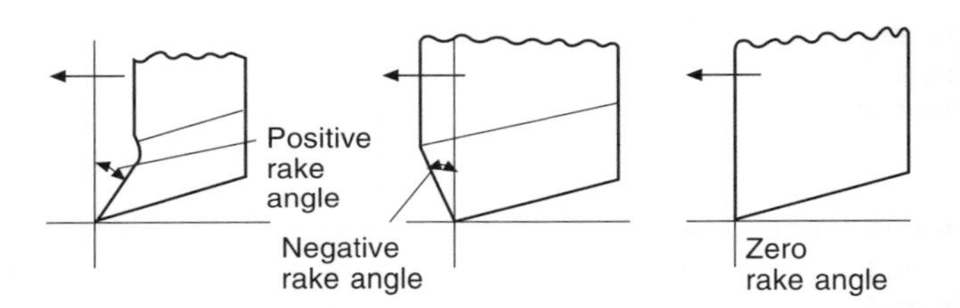

Fig. R.1 Rake angles

expansion of the steam in the engine to condenser pressure, and the condensation of the expanded, exhaust steam to its initial point.

Rankine scale (*sci*) An obsolescent temperature scale that is the absolute version of the Fahrenheit scale. As absolute zero is – 460°F, to convert temperatures expressed in degrees Fahrenheit to Rankine it is necessary to add 460 to the temperature in degrees Fahrenheit.

ratio (*maths*) The ratio of one number to another is a fraction, and is the number of times one quantity is contained in another quantity *of the same kind.* Since the numerator and denominator of the fraction have the same units, these cancel out and the ratio has no units. Thus the ratio of 25 N to 75 N is 1/3 and this ratio has no units.

reamer (*mech*) A tool for sizing an existing hole and improving its roundness and finish. Since reaming is only a finishing operation, a hole must first be drilled slightly under size, leaving just enough metal for the reaming operation to 'clean up'. A reamer will always follow the axis of the previous hole, it cannot correct positional inaccuracies. Positional inaccuracies can only be corrected by single-point boring.

reciprocals (*maths*) The reciprocal of a number is given when its index is –1, and the value of the reciprocal of a number becomes a fraction in which the numerator is 1 and the denominator is the number itself. Thus the reciprocal of 5 is 1/5 or 0.2 when expressed as a decimal fraction.

rectangle (*maths*) A quadrilateral in which all four angles are right angles, opposite sides are parallel and equal in length, the diagonals are equal in length and bisect each other.

rectification Process of converting *alternating current* to *direct current.*

rectifier Circuit that converts *alternating current* to *direct current* by use of a diode or a combination of diodes. See: *diode*; *full-wave rectifier*; *half-wave rectifier.*

recursive (*maths*) A type of system model, which estimates the current system output using historical data samples.

redundant member (*sci*) A member of a framed structure with zero force acting on it and which provides no support to the loaded structure.

Redwood second (*sci*) Unit of viscosity derived from the Redwood viscometer. See: *viscometry.*

refrigerant (*sci*) Liquid used in the refrigeration cycle. Common refrigerants are ammonia and Freon-22.

refrigerator (*sci*) (or heat pump) A closed thermodynamic cycle designed to absorb heat at a lower temperature and reject it at a higher temperature. The liquid refrigerant passes to an evaporator where it is evaporated and heat is transferred to the refrigerant at constant pressure. The refrigerant is then compressed in order to increase its pressure and temperature, and then passes to a condenser where heat is transferred to the atmosphere as the refrigerant is condensed at a constant pressure. The condensed refrigerant finally passes through an expansion valve where it is restored to its original pressure and temperature before passing to the evaporator again for the cycle to be repeated.

registers (*comp*) Storage place inside microprocessor for temporary storage of data.

regression (*maths*) Where perfect correlation or near-perfect

correlation does not exist, regression analysis is used to draw the line of 'best fit' through the coordinates of a graph. For the line to represent 'best fit', the deviation of all the coordinates from the line should be minimal.
regulated power supply See: *stabilized power supply.*
relative address (*comp*) The address of memory location in a computer that is specified relative to the program content instead of being specified as a fixed number. See: *absolute address.*
relative density (*sci*) The density of a substance compared to the density of pure water at 4°C. Hence

$$\text{relative density} = \frac{\text{density of the substance}}{\text{density of pure water at 4°C}}$$

or

$$\text{relative density} = \frac{\text{mass of the substance}}{\text{mass of an equal volume of pure water at 4°C}}$$

Since the above expressions are ratios there are no units.
relative frequency (*stats*) The relative frequency with which any member in a set occurs is given by the ratio: (frequency of the member)/(total frequency of all the members of that set). By multiplying by 100, this ratio can also be expressed as a percentage, when it becomes known as the percentage relative frequency.
reliability (*sci*) The probability that a system will operate as designed when required.
reluctance (*elec*) Reluctance can be considered as the 'resistance' of a magnetic circuit to the presence of magnetic flux. It is the quantity of a magnetic circuit that limits the flux produced by a given *m.m.f.* The quantity symbol is *S*, and the unit ampere-turns/weber (At/Wb). Air and non-magnetic materials such as copper and aluminium have a high reluctance, whilst the ferromag-netic

materials have a lower reluctance. For any magnetic circuit, reluctance can be defined as the ratio of the applied magnetomotive force to the flux in that circuit, expressed as:

$$\text{reluctance } (S) = \frac{\text{magnetomotive force } (F)}{\text{magnetic flux } (\Phi)}$$

where the unit of magnetomotive force (m.m.f.) is the ampere-turn (At), the unit of magnetic flux is the weber (Wb), and the unit of reluctance is the ratio ampere-turns per weber (At/Wb) or the henry (H). Reluctance is also the reciprocal of permeance. See: *permeance*.

resilience (*sci*) Elastic strain energy stored in a material when work is done to strain that material, provided the *elastic limit* is not exceeded.

$$\text{Resilience} = \frac{\sigma^2 V}{2E}$$

where σ is maximum direct stress, V is the volume of the material and E is *Young's modulus of elasticity* of the material. Torsional resilience is the energy stored when work is done to twist a shaft.

$$\text{Torsional resilience} = \frac{\tau^2 V}{2G}$$

where τ is the maximum shear stress and G is the *modulus of rigidity* of the material.

resistance (*elec*) Measure of the opposition to the flow of electric charge when an e.m.f. is applied; quantity symbol R; unit the ohm (Ω). For *d.c.* circuits, $R = V/I$. See also: *Ohm's law*.

resistance sensing element (*sci*) Sensing element for electronic instrumentation in which electrical resistance is an analogue of the measured quantity. For example, copper may be used for the sensing element of a resistance thermometer, since any change in resistance is directly proportional to the corresponding change in temperature. Another example is a strain gauge which converts a change in strain to a corresponding change in resistance when the element is distorted.

resistivity (*elec*) **1** The resistance offered by a unit cube of a given material between opposite faces. **2** The intrinsic property of a conductor which gives resistance in terms of its dimensions; $R = \rho l/A$, where R = resistance of conductor in ohms, l = its length in metres, A = its uniform cross-sectional area in m^2 and ρ = its resistivity in ohm metres.

resolution (*comp*) **1** The number of levels of a *digital-to-analogue converter*; e.g. 8 binary inputs gives $2^8 = 256$, and it has a resolution of 1 in 256 or 0.39%. **2** Number of elements per unit length of computer equipment—display, scanner, printer, etc.—describing the image sharpness and expressed as dots per inch (dpi).

resolution of forces (*sci*) The separation of a vector into two components, usually a horizontal and a vertical component for ease of calculation, i.e. $R \cos\theta$ and $R \sin\theta$ where R is the magnitude of the vector and θ describes the direction, as shown in Fig. R.2.

resonance (*elec*) For all a.c. circuits possessing resistance (R), capacitance (C) and inductance (L), there is a particular frequency when the voltage across the circuit and the current supplied to the circuit are *in phase*. This frequency is known as the *resonant frequency* and the circuit is said to be *resonant*. The phenomenon of resonance is exploited in the tuning of radio and television receivers. There are two types of resonant circuits. (1) Series resonance where the circuit impedance $Z = R$ and where the impedance is at a minimum at the resonant frequency (hence it is called an acceptor circuit). (2) Parallel resonance where the impedance is at a maximum at the resonant frequency (hence it is called a rejector circuit).

resultant (*sci*) A single vector that has the same effect as multiple vectors acting together through the same point.

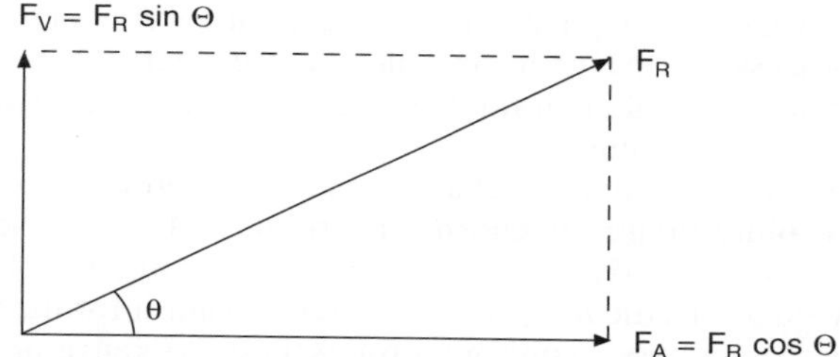

(a) Resolution of forces when F_V and F_H are at right angles

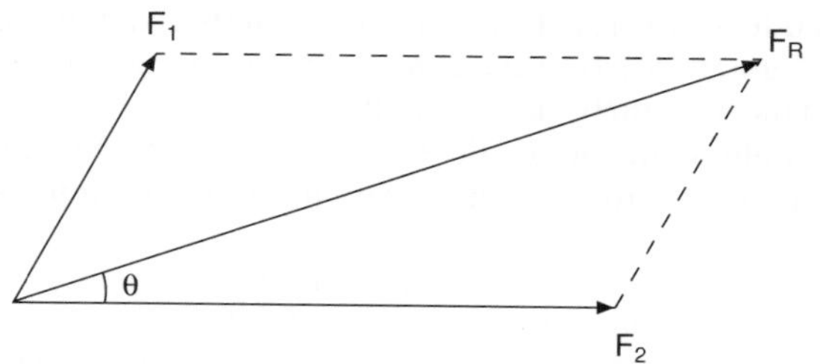

(b) Resolution of forces when F_1 and F_2 are not at right angles.
(i) Draw the parallelogram of forces and scale the drawing OR
(ii) Use sine and cosine formula to calculate the value of F_R.

Fig. R.2 Resolution of Forces

retardation (*sci*) For a body in motion, retardation is the rate of slowing down, i.e. negative acceleration. See: *acceleration*.
reverse engineering (*mech*) The dismantling and analysis of a product to investigate its design and manufacture, usually carried out by a competitor's company.
reversible process (*sci*) A thermodynamic ideal process consisting of a fluid that undergoes a process and can then be taken back through all the stages of the process in reverse order

and reach the original state of system and surroundings. A reversible process can never happen in practice owing to losses such as friction, churning of fluid, etc., making the process irreversible. It is often convenient to assume some processes are reversible to simplify calculations and the analysis of problems.

revolute (angular) coordinate robot (*CAE*) These have three rotary axes and no linear axes, as shown in Fig. R.3. The revolute, angular or articulated robot (different names for the same thing) has two rotary joints and a rotary base. Its range of movements closely resembles that of the human arm. Having three rotary encoders, its resolution is low compared with other robot configurations. However, the revolute configuration is the most popular for small- and medium-sized robots.

Reynold's number (*sci*) A dimensionless number that describes the type of flow of a liquid passing through a pipe, symbol R_e:

$$R_e = \frac{\rho v d}{\eta}$$

where ρ is the density of the liquid in kg/m^3, v is the fluid velocity in m/s, d is the pipe diameter in m, and η is the fluid viscosity in $N\ s/m^2$ or kg/m s. Laminar flow occurs when the Reynold's number is below 2000. The transition to turbulent flow occurs when the Reynold's number lies between 2000 and 2500, with the flow being entirely turbulent for numbers above 2500.

rhombus (*maths*) A special case of the parallelogram where all four sides are of equal length, the diagonals bisect the corner angles, and the diagonals bisect each other at right angles. See: *parallelogram*.

right-angled triangle (*maths*) See: *triangle*.

rigidity (modulus of) (*sci*) See: *modulus of rigidity*.

riser (*mech*) Channel from a *mould* for carrying dirt, slag and sand away from the *casting*. As the mould is filled through the

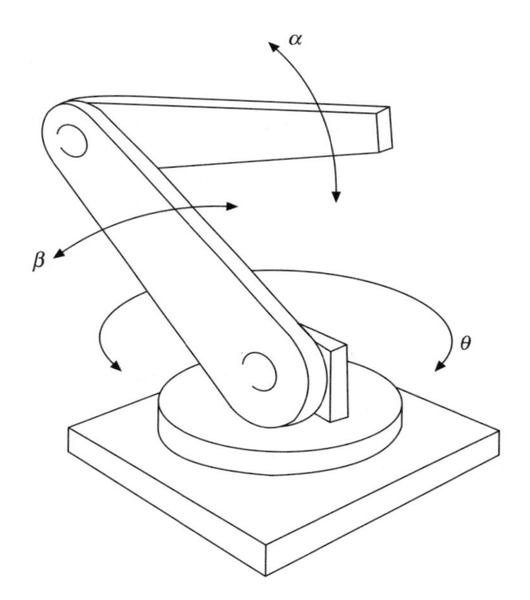

Fig. R.3 Revolute Coordinate Robot

runner, molten metal flows up into the riser, indicating when the mould is full, carrying away waste products, and then assisting in feeding the casting as it cools and contracts. It also allows a passage through which the air in the mould cavity can be displaced as the molten metal enters the mould. Complex moulds require more than one riser strategically placed by a skilled moulder.

rivet (*mech*) A component for making a permanent joint between two plates. Rivets are made of low-carbon steel and are annealed so that they are soft. They consist of a shank and a shaped head. The rivet is inserted into a hole drilled into the two overlapping plates to be joined and the shank end is 'closed' by hammering or by machine so that the rivet swells to fill the hole and to form a second head. The process can be carried out with red-hot rivets (hot-riveting) or at room temperature (cold-riveting). Hot-riveting is used for the larger sizes of rivet to ease the force required to close the rivet. Also, as the rivet cools it shrinks and pulls the joint tightly together. Welding has now largely replaced riveting.

robot (*CAE*) See: *industrial robot.*

robot end effectors (*CAE*) These are special *grippers* and *tools* that are attached to the wrist assembly to suit the job that the robot is being called upon to perform. They may be considered as the 'hand and fingers' of the robot. *Grippers* can be mechanical, magnetic or suction. *Tools* can be screwdrivers, nut runners, assembly tools, welding tools, inspection probes, power tools, spray-paint guns, etc.

robot-programming methods (*CAE*) Robot control systems share a number of hardware and software similarities with computer numerically controlled (CNC) machine tools. They use the same linear and rotary encoders, the same stepper and servo drives, they can have open- or closed-loop control systems and are controlled by a dedicated computer. However, the method of programming can be substantially different to those used with CNC machines. Robot-programming languages will not be considered as they are so numerous and vary according to the make and type of robot. See: '*lead-through*' *programming*; '*drive-through*' *programming*; '*off-line*' *programming*.

Rockwell hardness (*matls*) A test used to measure the hardness of metal by applying a load to a steel ball-bearing or a diamond cone to create an indentation. The hardness is a measure of the difference in penetration of the indenter between a fixed minor load and a specified major load. The hardness is read directly from the instrument dial.

roller bearing (*mech*) A rotary anti-friction bearing consisting of a number of parallel or tapered hardened-steel rollers that roll between a hard steel inner race and a hard-steel outer race. The rollers are evenly spaced within the races by means of a cage. Roller bearings can support heavier loads than ball-bearings. See: *ball-bearings*; *needle roller bearings*.

ROM (*comp*) Read-only memory, a non-volatile memory of a computer, i.e. one that remains intact when the power is switched off, and hence used for permanent storage of data, e.g. software.

root-mean-square (*elec*) (Abb: r.m.s.) The equivalent steady value of a varying quantity, calculated by squaring the ordinates of a complete cycle, summing the squared ordinates and then taking an average. For alternating currents where the waveform is sinusoidal, $I_{rms} = I_m/\sqrt{2}$ where I_m is the maximum value of the current. The r.m.s. value of current is the equivalent d.c. value that would dissipate the same power with the same circuit, and is a means of comparing the effect of an alternating current value with a direct current value.

roots (*maths*) The root of a number is found when it is raised to a power that is a proper fraction. Thus the square root of $16 = 16^{1/2} = 4$, or $\sqrt{16} = 4$, since $4 \times 4 = 16$. Similarly, the cube root of $27 = 27^{1/3} = 3$, since $3 \times 3 \times 3 = 27$. Further roots, such as $32^{1/5}$, is referred to as the fifth root of 32. Note that square roots always have two answers since $2 \times 2 = 4$ and $-2 \times -2 = 4$, therefore $\sqrt{4} = \pm 2$.

rotameter (*sci*) Instrument for measuring the rate of flow of *fluids*.

rotation (*CAE*) This is a programming facility, available with some controllers, that allows the whole coordinate system for a milling or drilling operation to be rotated through a stated angle as shown in Fig. R.4. A subroutine is written to machine the cavity in the first position and then the rotation facility is used to machine the remaining cavities in the subsequent positions. Assuming the controller acknowledges the ISO rotation address code G73 and the H address is used to store the rotation angle, then the program would be:

N100 call subroutine to machine cavity	Causes cavity to be milled at position 1.
N110 G73 H45	Rotates the coordinate system by 45° from angle zero.
N120 call subroutine to machine cavity	Causes cavity to be milled at position 2.

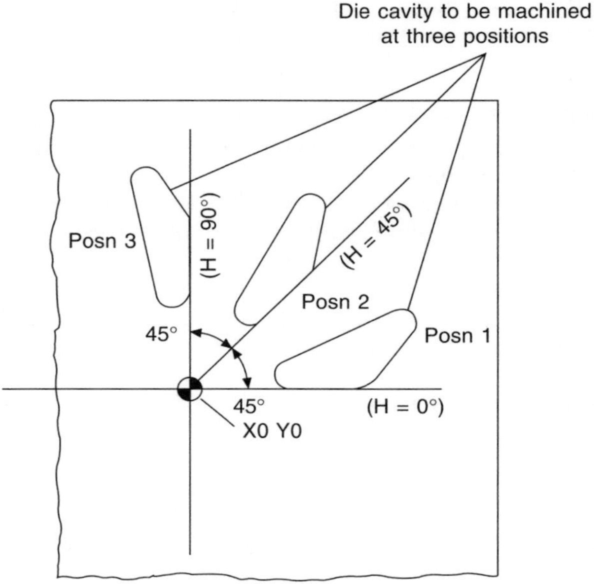

Fig. R.4 Rotation

N130 G73 H90	Rotates the coordinate system by 90° from angle zero.
N140 call subroutine to machine cavity	Causes cavity to be milled at position 3.
N150 G73 H0	Sets the coordinate system back to normal (0°)

rotational momentum (*sci*) See: *momentum.*
rotor Rotating part of a machine such as an electric motor, pump, turbine, etc., the *stator* being the stationary part.
RTL Resistor transistor logic, an early form of integrated circuit technology.
runner (*mech*) A channel down which molten metal is poured into a mould.
rusting (*matls*) The wet corrosion of iron or steel to form iron hydroxide. Rusting occurs in the presence of water and oxygen. The extent and rate of rusting increases with a higher level of impurities in the metal and dissolved salts in the water. Not to be confused with dry oxidation products such as mill scale. See: *mill scale.*

S

sacrificial anode (*matls*) A block of metal that is fitted to an underwater structure so the block of metal acts as an anode in any electrolytic reaction and is itself eaten away whilst preventing corrosion to the structure. Since the anode is eaten away it is said to be *sacrificial*. See: *cathodic protection*.
saddle (*mech*) That component of the carriage of a centre lathe that spans the bed and is supported on, and guided by, the main longitudinal slideways (shears). In turn, it carries the cross-slide, compound (top) slide and toolpost.
sagging (*mech*) The bending of a beam where the ends of the beam tend to move up and the middle section of the length tends to move down. By convention, this is usually considered as a positive deflection convention in beam-bending theory. See: *hogging*.
salt-bath furnace (*matls*) An electrically or gas-heated furnace in which the work (charge) is immersed in molten salts. This type of furnace is used for heat treatment processes since the high heat capacity of the molten salt ensures uniform heating and enables accurate, automatic temperature control to be easily achieved. Since it is immersed in the molten salt, the work is protected from atmospheric oxidation (scaling). The salts may be neutral for most heat treatment processes, but carburizing salts are used for case-hardening processes. See: *carburizing*; *case hardening*.
sample (*stats*) A selection of *members* from a *population*.
sand-casting (*mech*) See: *casting*.
saturated liquid (*sci*) A liquid that has just reached its saturation temperature but is still all liquid and no vapour, having a dryness fraction of 0.

saturated steam (*sci*) Steam that is at the same temperature as the water from which it is formed. It will contain water particles. See: *superheated steam.*

saturated vapour (*sci*) A vapour that is sufficiently concentrated to coexist in equilibrium with the liquid from which it was formed.

saturation temperature (*sci*) The temperature at which the state change from a liquid to a gas takes place at a specified pressure, i.e. the saturation pressure.

scalar quantity (*maths*) A quantity in which magnitude only is stated and direction is not applicable, e.g. volume, voltage, current, etc. See: *vector quantity.*

scalene triangle (*maths*) See: *triangle.*

scaling (*CAE*) The scaling feature allows *x*- and *y*-coordinates in milling and *x*- and *z*-coordinates in turning to be increased or decreased by a scaling factor from their stated values in the program.

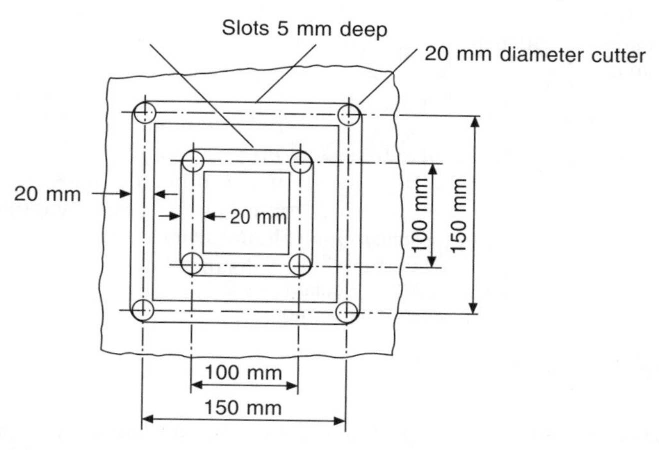

Fig. S.1 Scaling

For example, the slots to be milled as shown in Fig. S.1 do not have to be programmed separately. A macro is written for the inner slot and called up twice, the first time without the scaling facility and the second time with the scaling facility active, with a scaling factor of 1.5 (150/100). Alternatively, the subroutine could be written for the outer slot and scaled 2/3 for the inner slot.

scatter (*stats*) See: *standard deviation*.

scraper (*mech*) A hand tool for finishing metal surfaces by removing flatness errors caused by local high spots. The surface being finished must be reasonably flat to start with. Flat surfaces are finished with a flat scraper and internal cylindrical surfaces are finished with a half-round scraper (see Fig. S.2). See also: *flatness*.

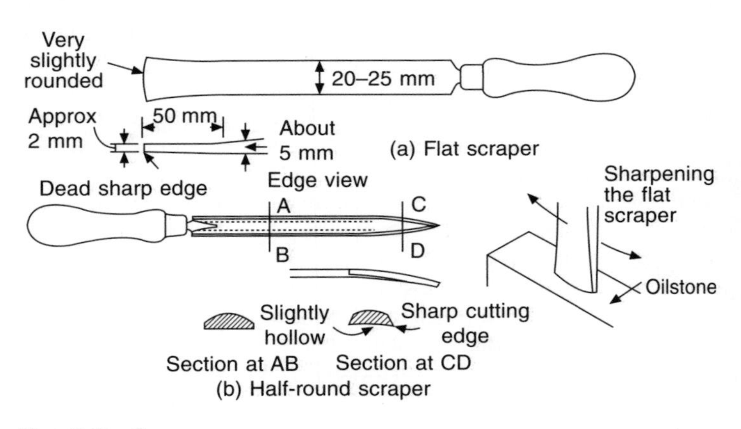

Fig. S.2 Scrapers

screw thread (*mech*) In principle, a screw thread can be considered as a helical ridge, usually of V-section, formed or generated on a cylindrical base. In action it is effectively the same as a wedge shape driven between two surfaces but with the wedge wound onto the cylinder and thus providing longitudinal thrust when the cylinder is rotated within a hole with a mating internal thread. Various screw-thread forms are available for different duties, and the dimensions of screw threads are listed in standard screw-thread tables.

scriber (*mech*) A marking-out instrument, with a fine, sharp, hardened and tempered point, for scribing lines on a workpiece.

scribing block (or surface gauge) (*mech*) A marking-out instrument comprising a *scriber* clamped to a mast which is supported on a base. The position of the scriber on the mast is adjustable for coarse setting and the rake of the mast is adjustable for fine setting. This instrument is used for scribing lines parallel to the base as it is moved along a *marking-out table* which provides a common datum surface for the workpiece and the scribing block. It is also used to test parallelism and alignment of holes. Since the accuracy of lines marked out with the use of a scribbling block is limited by the accuracy of the rule to which it is set, for more precise marking out, a vernier height gauge should be used. See: *vernier height gauge*.

seal (*mech*) (1) Static seals are sandwiched between two component faces to take up surface contours, e.g. a synthetic rubber 'O' ring. See also: *gasket*. (2) Dynamic seals are annular shaped components that fit into the stator of a machine through which the rotor passes, preventing leakage of fluid past the components in relative rotary motion. The seal comprises a sleeve and flexible web that stays concentric with the shaft, and a seal lip with a knife-edge that is kept in contact with the shaft by means of a *garter* spring. For example, an oil seal is used on either end of an engine crankshaft as it passes through the engine casing to prevent oil leaks.

second moment of area (*sci*) The product of the magnitude of an area and the distance of the centre of the area to the axis squared, e.g. as used in the *bending equation*.

segar cone (*matls*) Moulded cone-shaped material designed to collapse and melt at a specific temperature. Furnace temperatures can be monitored by the use of cones ranging from just below to just above the required temperature. The furnace temperature is established by observing which cones have collapsed.
self-inductance (*elec*) The property of a circuit that causes an e.m.f. to be induced when the current through it changes; quantity symbol L, unit the henry (H). A circuit has a self-inductance of one henry when an e.m.f. of one volt is induced when the current through it changes by one ampere,

$$E = -L \frac{\mathrm{d}I}{\mathrm{d}t}$$

where $\mathrm{d}I/\mathrm{d}t$ is the rate of change of current and E is the induced e.m.f.
sensible heat (*sci*) Heat energy that causes a temperature change of a body, $Q = mC\,\Delta\theta$ where Q is the heat transfer, m is the mass of the body, C is the specific heat capacity and $\Delta\theta$ is the temperature change.
sensing element (*sci*) The element of an instrumentation system that converts the variable to be measured into a more suitable corresponding signal for conditioning and display, e.g. in a spring balance the force applied is proportional to the deflection of the spring, and so the force signal is converted into a corresponding length signal.
sensitive feed (*mech*) This is a manual feed system used in drilling machines and which incorporates a rack and pinion mechanism. It gets its name from the fact that the operator can feel the resistance the workpiece material offers to the penetration of the drill and allows the feed force to be adjusted accordingly by the operator. This is particularly important when using very small diameter drills of low physical strength.
sensitivity (*sci*) The ratio of the change of magnitude in the

output of a system, to a given change of magnitude in the input to the system. If the change in the input is ΔQ_i and this produces a change in the output of ΔQ_0, then sensitivity = $\Delta Q_i / \Delta Q_0$.
sequential logic (*comp*) A logic circuit where the output (Q) is determined both by the state of the inputs and also by the previous values of the input. Therefore a sequential logic circuit has a memory. The basic building block of a sequential circuit is the *bistable*. This is a simple one-bit memory device. See: *bistable*.
series connection (*elec*) A method of connecting the elements of an electrical circuit in a simple loop so that the current flow is common to all the elements of the circuit, as shown in Fig. S.3. The e.m.f. across the circuit is equal to the sum of the potential differences across the individual circuit elements. If one circuit element fails, the circuit is broken and the current flow ceases in all the circuit elements. See: *parallel connection*.

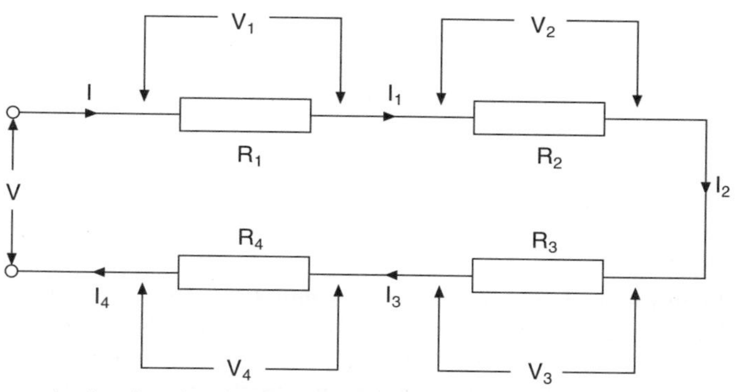

$I = I_1 = I_2 = I_3 = I_4$ (Current is constant)
$V = V_1 + V_2 + V_3 + V_4$
$R = R_1 + R_2 + R_3 + R_4$ where R = total circuit resistance.

Fig. S.3 Series Connection

set (*stats*) A grouping of related statistical data.
shaft power (*sci*) The power that a machine or engine produces at its output shaft, as measured by some type of dynamometer (torque measure). A friction or a hydraulic brake is one method of measuring the output torque and this test is commonly called *brake power*. The brake power or shaft power is always less than the indicated power owing to internal mechanical losses of the machine or engine.
shear force (*sci*) An applied load that consists of two forces that are equal and opposite, and parallel to each other but not acting in the same straight line. As the forces are not direct, there is a tendency for one part of the material to slide over another, as shown in Fig. S.4.

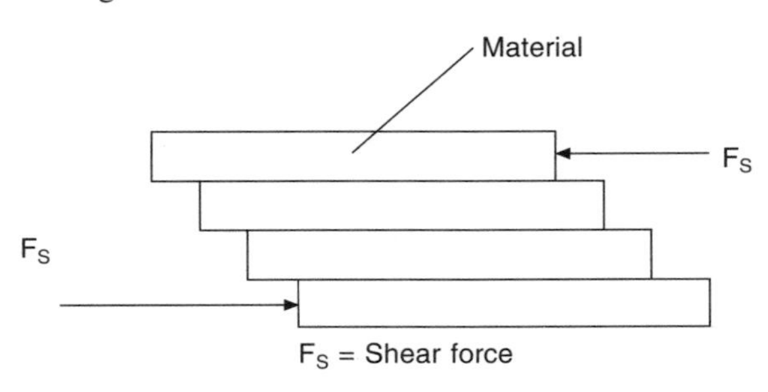

Fig. S.4 Shear Force

shear force diagram (*sci*) A graphical representation of shear force plotted against the length of a beam subject to bending; the horizontal axis represents the length of the beam and the vertical axis represents the shear force at that beam section, as shown in Fig. S.5.
shear strain (*sci*) A measure of the deformation of a material when a shear force is applied, being measured by the angular distortion of the faces perpendicular to the line of force; quantity

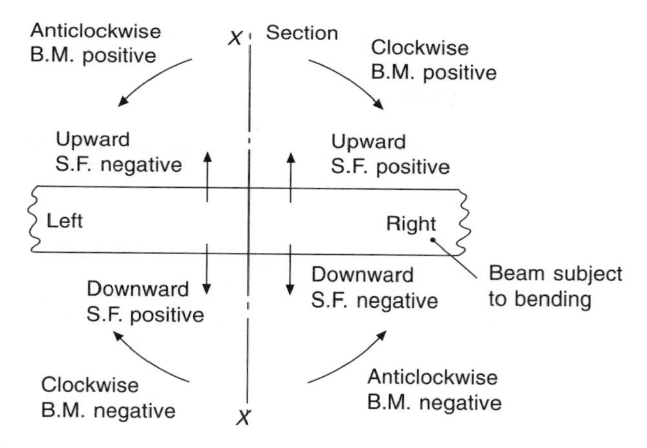

(a) Bending moment (B.M.) and shear force (S.F.) sign convention

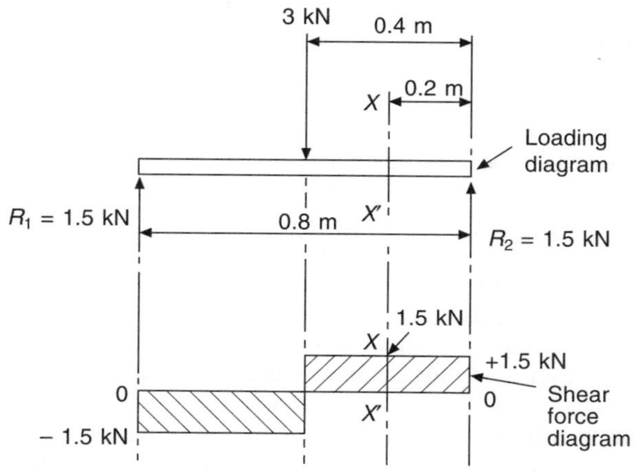

(b) Shear force diagram for a point load

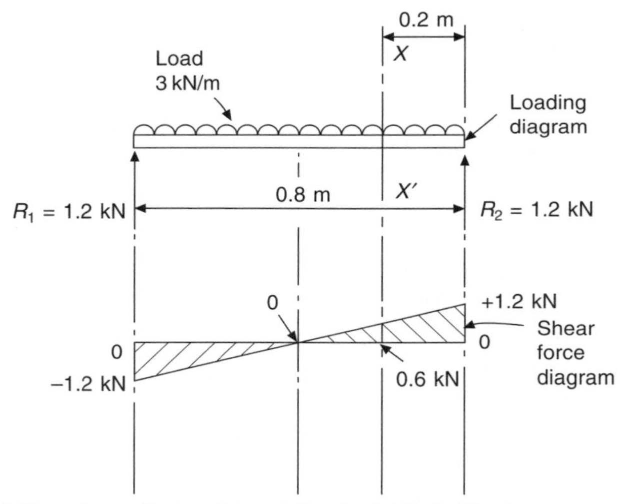

(c) Shear force diagram for a uniformly distributed load

Fig. S.5 Shear Force Diagram

symbol γ as shown in Fig. S.6. The distortion is strictly measured as the movement of the faces in the direction of the force divided by the distance between the faces and is equal to tan γ, γ being the angular distortion of the vertical faces; if the angle is small, then tan $\gamma = \gamma$ in radians.

shear stress (*sci*) The stress occurring when shear forces are applied to a material; shear stress $\tau = F/A$ where F is the shear force and A is the cross-sectional area parallel to the line of force.

shell moulding (*mech*) A moulding process in which the moulding sand is mixed with a binder so as to make a hard, strong shell that can be stored and handled in advance of

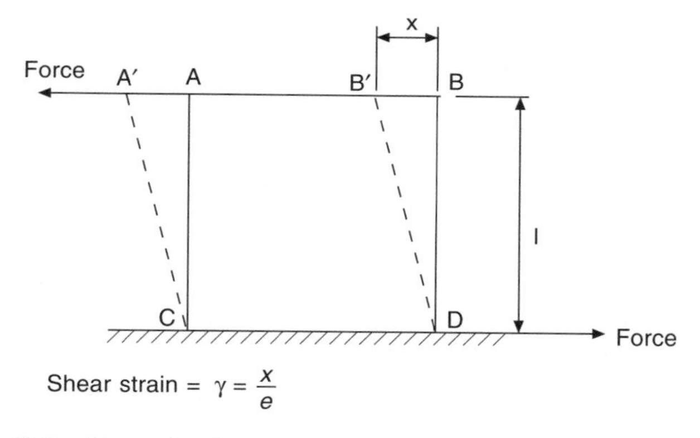

Fig. S.6 Shear Strain

requirement. Figure S.7 shows the steps in making a shell mould. The shell is made in two parts so that it can be stripped from the mould and then assembled using an adhesive ready for the molten metal to be poured. When a phenolic resin binder is used the pattern is heated in an oven to a temperature of 200–250°C. The pattern is sprayed with a release agent and clamped to a 'dump box'. The dump box is then inverted. The sand–resin mixture falls onto the heated half-pattern and forms a soft 'biscuit' about 6 mm thick. The dump box is then rotated to its original position so that the surplus sand–resin mixture falls away from the pattern ready for reuse. The pattern and biscuit are transferred to an oven for about 2–5 minutes to cure the shell, after which it is stripped from the pattern ready for assembly. Because of the cost of the equipment and metal patterns, this process is more suitable for repeated production of castings. An alternative to the use of phenolic resin as a binder is to use silicon compounds that are set off by a reaction with carbon dioxide gas.

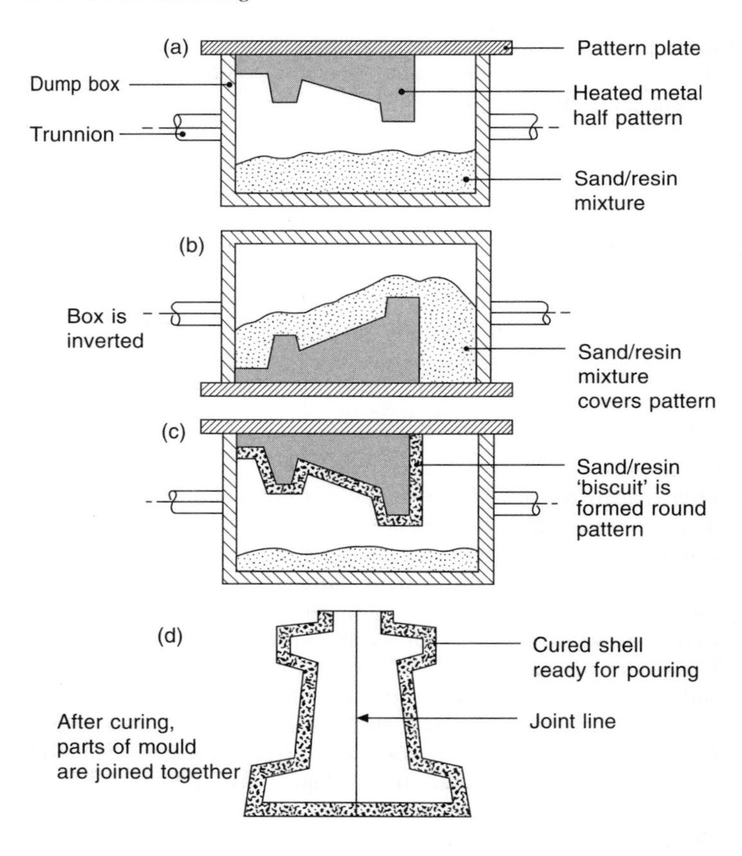

Fig. S.7 Making a Shell mould

sherardizing (*matls*) See: *cementation processes.*
short circuit (*elec*) The connection of a *conductor* of near *zero impedence* between two points on a circuit. This reduces the potential difference to zero quickly and can result in damage owing to the high currents that flow unless overcurrent protection is present. See: *overcurrent protection.*
shunt (*elec*) **1** Term used for parallel connection of circuit elements, e.g. a shunt-wound d.c. motor or generator has the field winding wired in parallel with the armature windings. **2** Resistance connected in parallel in an electrical circuit. A common use is in instrumen-tation, for example an ammeter. By fitting a low-resistance conductor across the instrument, an extra path is provided to bypass current in excess of that required by the meter. Different shunts can be used and a switch fitted to select different ranges of current measurement.
siemen (*elec*) Units of *conductance*, S.
Siemens process (*matls*) Obsolescent open-hearth steel-making process named after the inventor. See: *open hearth.*
signal conditioner (*elec*) Component of instrumentation system, that changes the signal from the sensing element to a more suitable form for a display, recorder or control system.
silver solder (*matls*) A hard-soldering process graded between soft-soldering and brazing in terms of the strength of the joint and the temperatures involved. The solder is so called because it contains a high percentage of the precious metal silver. In the hands of a skilled worker, silver soldering produces neat joints that have a substantially higher strength and ductility than soft-soldered joints.
simple harmonic motion (*sci*) A precise periodic motion to which many types of motion in engineering may be approximated for purposes of analysis, e.g. the motion of a reciprocating piston, a vibrating mass on a spring, the swing of a simple pendulum. Simple harmonic motion is defined as the motion of a body whose

acceleration is always towards a fixed point in its path and is proportional to its displacement from that point.

simply supported (*sci*) An idealistic support presumed for problems involving the loading of structures and beams, where the total force exerted by the body or structure on the ground is considered to act through two *knife-edge supports* that in no way impede any flexure of the beam under load conditions.

Simpson's rule (*maths*) With reference to Fig. S.8, the approximate area under the curve AB can be determined by dividing the base OX into an **even** number of intervals of width x (the greater the number of intervals, the greater will be the accuracy) and measuring the corresponding ordinates y_1, y_2, y_3, etc. The area of the figure is obtained as follows:

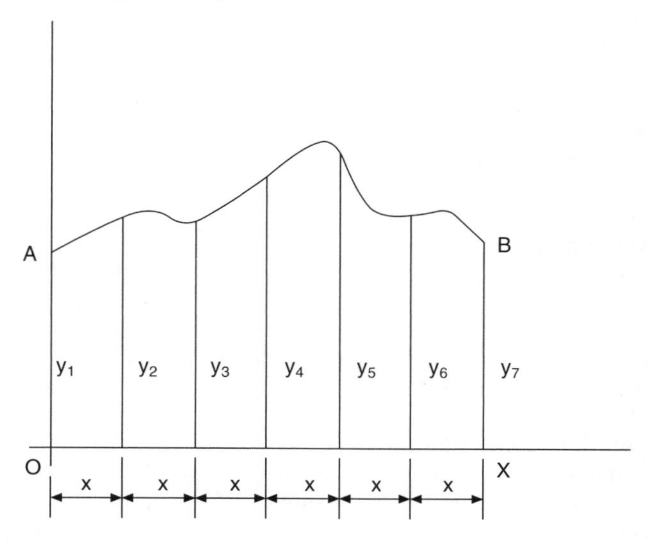

Fig. S.8 Simpson's Rule

$$\text{area} = x/3[(y_1 + y_7) + 4(y_2 + y_4 + y_6) + 2(y_3 + y_5)]$$

In general terms the rule can be expressed as: *area = 1/3 width of interval [(sum of first and last ordinates) + 4(sum of even ordinates) + 2(sum of odd ordinates)].*

simultaneous equations (*maths*) Equations expressed in terms of unknown variables that can be solved if the total number of equations is greater than the number of unknown variables.

sine bar (*mech*) A device as shown in Fig. S.9(a), used for accurately measuring angles. The sine bar, the slip gauges and the surface plate form a right-angled triangle in which the sine bar forms the hypotenuse and the slip gauges the 'opposite side' to the angle being set or measured, as shown in Fig. S.9(b). The distance between the axes of the contact rollers is the length of the hypotenuse and is normally 125 mm or 250 mm. The height of the slip gauge stack is calculated using trigonometry. The component whose angle is to be checked is placed on the sine bar as shown in Fig. S.9(c). If the angle being measured is correct the dial test indicator (DTI) should show a constant reading as it is slid along the surface of the work. Any deviation is the error.

single-phase supply Domestic two-wire (live and neutral) a.c. supply consisting of a single sinusoidal waveform.

single-point tools (*matls*) Cutting tools with a single point such as those used on shaping machines, planing machines and lathes. Such tools have the advantage that they are easy to sharpen and refurbish using a double-ended, off-hand grinding machine, useful for jobbing work. Figure S.10 shows some typical lathe tools and their applications. See: *pre-set tooling; qualified tooling.*

SI system (*sci*) Système International d'Unités; system of metric units of quantity comprising *base* units, *derived* units, *multiples* and *submultiples*, each with an agreed symbol. See: *Appendix 4 metric units.*

slag (*matls*) **1** During the smelting of iron ore, slag is formed by

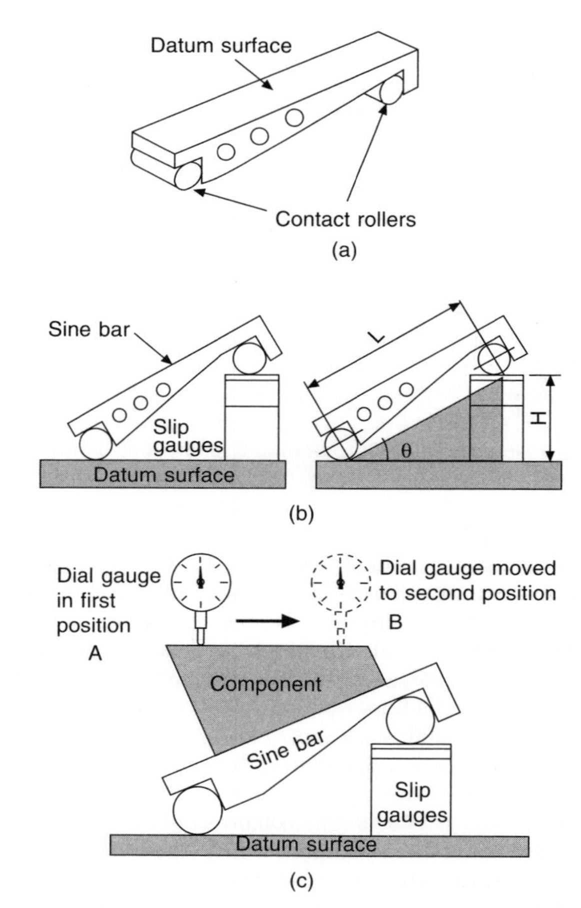

Fig. S.9 The Sine Bar: (a) design; (b) principles; (c) use of the sine bar

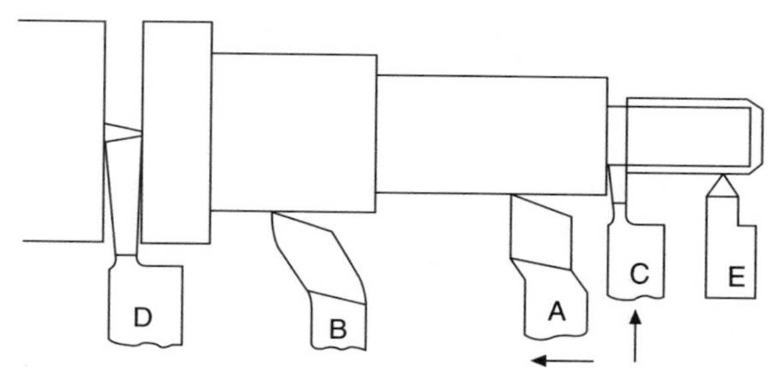

A = Knife tool
B = Turning and facing tool
C = Undercutting tool
D = Parting-off tool
E = Screw-cutting tool (55° or 60°)

Fig. S.10 Single-point Tool Applications

the chemical reaction of the earthy impurities in the ore with the limestone (calcium carbonate) that is added in the charge. The molten slag floats on top of the molten iron and this unwanted by-product is run off before the iron is poured. The solidified slag is crushed and used as a road-making material. A similar slag is also formed during the basic steel-making process and is used as a fertilizer as it has a high phosphate content. **2** During electric arc welding any impurities, such as atmospheric oxidation products combine with the flux coating of the electrode to form a protective layer of fusible slag over the weld.

slip gauges (*mech*) Rectangular, hard-steel blocks supplied in sets of graded sizes. Their gauging surfaces are made to very high standards of dimensional accuracy, parallelism and surface finish. This enables two or more blocks to be *wrung* together when building up the required gauge dimension. They are used as primary standards of length in engineering workshops and inspection departments.

slip rings (*elec*) Copper rings connected to the rotor conductors and mounted on the rotor shaft of an a.c. motor or generator so that they turn with the rotor. Stationary carbon brushes press on the rings so that electrical currents can be delivered to or retrieved from the rotor conductors.

slot drill (*mech*) Also called a cotter-mill. This is a two- or three-flute milling cutter with teeth on the side and the end which can be used for cutting slots in metal without the need for a starting hole. See: *milling cutters.*

smoothing (*elec*) Compared with the uniform d.c. output of a battery, rectified a.c. has a pronounced ripple. Before rectified a.c. can be used to power most electronic devices, the ripple must be removed by a smoothing circuit. Such a circuit consists of a capacitor and choke or a capacitor and resistor filter, as shown in Fig. S.11.

soft-solder (*matls*) An alloy of tin and lead with a little antimony used for soft-soldering. Solders of different compositions are available and are selected according to the requirements of the joint being made. For example, Grade A solder has 65% tin, 34.4% lead and 0.6% antimony; its melting range is 183–185°C. Because of its short melting range and low electrical resistance it is widely used for making joints in electrical and electronic equipment. Grade K solder (*Tinman's solder*) has a composition of 60% tin, 39.5% lead and 0.5% antimony; its melting range is 183–188°C. This is a good general-purpose solder. Grade J solder (Plumber's solder) has a composition of 30% tin, 69.7% lead and 0.3% antimony, its melting range is 183–255°C. Because of its wide melting range this solder becomes 'pasty' before it melts and can be used for plumber's *wiped* joints.

soft-soldering (*mech*) A method of joining metals such as steel, copper and brass, by use of a jointing alloy called solder that

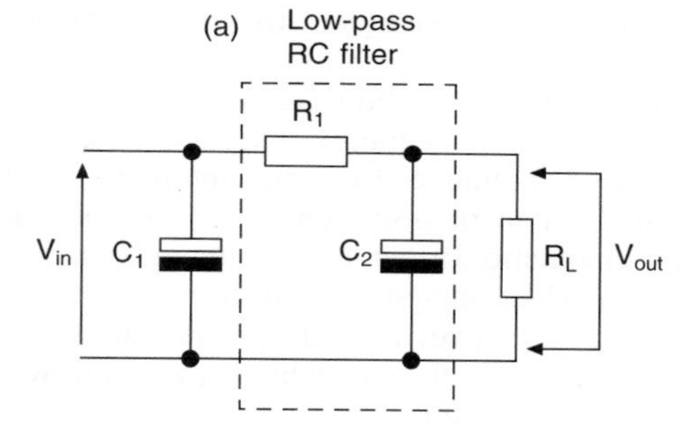

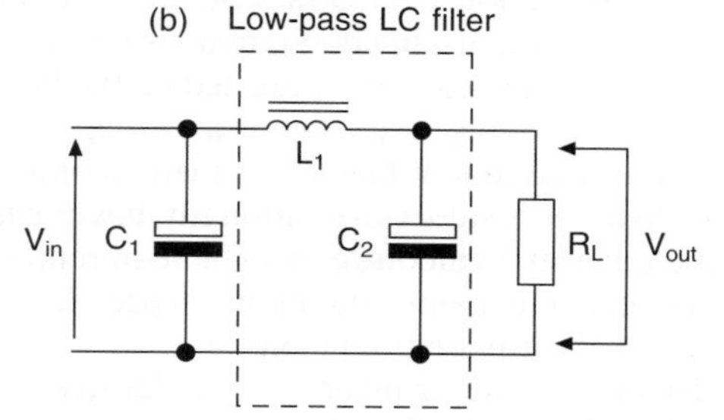

C_1 = Reservoir capacitor, C_2 = Smoothing capacitor
R_1 = Resistor, R_L = Load, L_1 = Inductor (choke).

Fig. S.11 Smoothing Circuits

'bonds' to the unmelted surfaces of the parent metals being joined in the presence of a suitable flux upon the application of heat. The solder melts and forms an amalgam with the surfaces of the parent metal. The flux chemically cleans the joint surfaces and prevents their oxidation whilst being heated. The flux also 'wets' the joint surfaces and helps the molten solder to run into the joint. See: *soft-solder*.

solar cells (*elec*) These are solid state devices that convert light energy directly into electrical energy. They are increasingly used to power a range of appliances from calculators to satellites.

solenoid (*elec*) A coil of insulated wire wound on a hollow, cylindrical non-magnetic former for producing a magnetic flux field. To make an electromagnet, an iron core can be inserted into the solenoid. Alternatively, a soft-iron plunger that is pulled into the solenoid parallel to the axis when a current flows through the wire coil can be used to operate circuit-breakers and valves.

space diagram (*sci*) A diagram of a system of vectors acting on a body. Used in conjunction with Bow's notation, it is drawn to simplify and clarify problems; the diagram does not show objects in contact with a body, just the forces exerted on the body and by the body on its surroundings. See: *Bow's notation*.

specific (*sci*) A term used to relate the properties of a substance to unit mass of that substance. Conventionally lower-case letters are used for the quantity symbols, e.g. volume is represented by V in the SI system and measured in m^3; specific volume is represented by v and measured in m^3/kg.

specific gravity (*sci*) Old term for relative density.

specific heat capacity (*sci*) See: *heat capacity*.

speed (*sci*) The scalar component of *velocity*, i.e. the distance covered by a body in motion per unit of time, the direction not being specified.

spelter (*matls*) A brass alloy used as the filler material when brazing. As for solders, there are various compositions depending on the particular application. For example, a British Standard 1845 type 8 alloy has 49–51% copper and the balance zinc. A boron-based flux is used, whose melting range is 860–870°C. During the jointing process some of the zinc content vaporizes off leaving the joint alloy with a higher copper content. This substantially increases the strength of the joint. See: *brazing*.

spherical (polar) coordinate robot (*CAE*) These have one orthogonal linear sliding axis and two rotary axes, as shown in Fig. S.12. The arm can move in and out and tilt up and down on a horizontal pivot. The whole assembly can rotate about a vertical axis. The working envelope (volume) is a spherical shell. The resolution is limited by the use of two rotary encoders and is not constant but varies as the arm moves in or out. If this limited resolution of the wrist assembly can be tolerated, this type of

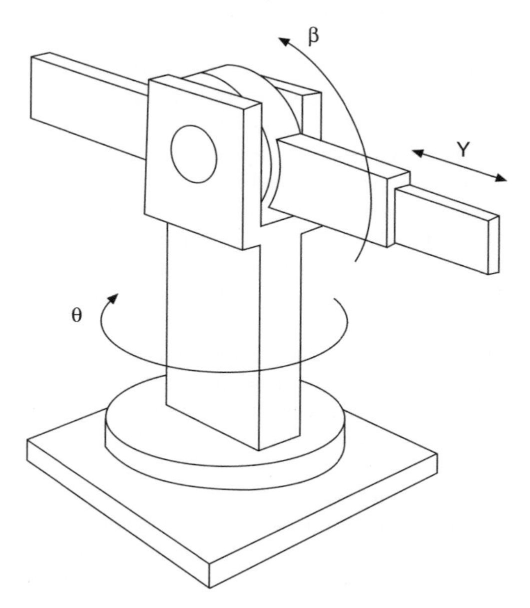

Fig. S.12 Spherical Coordinate Robot

robot configuration has the advantage of being quicker and more flexible than the high-resolution Cartesian robot.

spheroidal graphite (SG) cast iron (*matls*) Also known as *nodular iron, ductile iron, high-duty iron*, etc. The addition of magnesium or cerium to a normal grey cast iron causes the residual flake carbon to be redistributed uniformly throughout the mass of the metal as fine spheroids of carbon. Unlike the flake graphite of grey cast iron that leaves sharp corners as 'stress raisers', causing points of weakness, the carbon spheroids do not create stress concentrations to the same extent and this greatly enhances the strength, toughness and ductility of the castings. It also improves the resistance of the castings to fatigue failure.

spigot (*mech*) A raised, circular projection on the face of one component for location in a corresponding circular recess (register) in a mating component.

spindle (*mech*) **1** Generally, any shaft or axle in a mechanism. **2** The hollow shaft in the headstock of a lathe that supports, locates and drives the workholding devices such as chucks, centres, etc. **3** The hollow shaft of a milling machine that carries, locates and drives the cutter arbor or end mill chuck. The arbor or chuck is retained in position in the spindle nose by a draw bolt passing through the spindle.

spirit level (*mech*) Also called a bubble level. Cheap levels of limited accuracy consist of a glass tube (called a *vial*) bent into an arc and mounted transversely or longitudinally on a base. The tube is filled with spirit leaving a bubble of air to rise to the highest point of the tube. Precision levels used in engineering are fitted with vials having bores that are precision ground internally to a barrel shape since it is impossible to bend a tube to a sufficiently accurate curvature. Levels are used to test and/or set vertical and horizontal surfaces.

splash lubrication (*mech*) A method of applying lubrication by allowing machine components to repeatedly dip into a reservoir of oil and splash the lubricant over surrounding areas that require lubrication. For example, at least one of the gears in the headstock of a lathe dips into the oil reservoir and splashes oil over all the other gears and components in the gearbox.

spot facing (*mech*) A process for producing a shallow machined recess in the rough surfaces of a casting or forging in order to provide a flat seat for nuts and/or bolt heads. The cutter used is similar to a piloted counterbore but the diameter of the cutter is larger in proportion to the pilot that locates in a previously drilled bolt-hole.

spreadsheet Software program used for organizing, displaying and manipulating numerical data, with the display divided up into cells; the entry into each cell can be a value, formula or text. Formula entries can be recalculated automatically when new data is entered and statistical calculations and mathematical functions can be performed. Data may also be displayed in the form of different types of graphs and charts.

spring (*sci*) A mechanical device that can be strained in tension, compression or torsion to store energy, assuming the material from which it is made is not stressed beyond its elastic limit. If the deflection (x) is directly proportional to the applied compression force (F), then $F = kx$, where k is the spring stiffness and the constant of proportionality; the energy stored = $^1/_2 kx$.

spring balance (*sci*) An instrument that uses a spring to convert the applied force into a corresponding length, commonly used to measure weight with the deflection measured on a scale read-out as an analogue of the load.

spring stiffness (*sci*) See: *spring*.

spur gear (*mech*) Simple form of gearwheel with straight teeth parallel to the axis of the gear. Noisier and less smooth running than helical tooth gears but much easier and cheaper to manufacture.

square (*maths*) A special case of the rectangle in which all the sides are the same length. See: *rectangle*.

square thread (*mech*) A thread profile that is square-shaped with the depth of thread equal to half the pitch. It is not as strong as the corresponding V-form thread but has reduced friction when used for converting rotary motion into linear motion, as in a machine tool traverse screw or a screw-jack.

stabilized power supply (*elec*) A power supply (power pack) that maintains a constant d.c. output voltage despite variations in the a.c. supply voltage and/or the output current load. Also called a regulated power supply.

stainless steel (*matls*) A group of stain- and corrosion-resistant alloy steels all of which contain a high percentage of chromium. Some also contain a high percentage of nickel. The thin but dense film of chromium oxide that develops on the surface protects the steel from further corrosion. Unfortunately, these steels suffer from carbide precipitation, if cooled slowly through the temperature range of 650–800°C, which results in a marked reduction in strength. Where this is a result of fabrication by welding it is referred to as weld decay. Where it is the result of heat treatment or service environment it is referred to as temper brittleness. See: *weld decay*.

standard atmosphere (*sci*) **1** For general scientific purposes a standard atmosphere has a pressure of 101 325 N/m^2, which is equal to a 760-mm column of mercury at 0°C. **2** For meteorological purposes it is an internationally agreed theoretical state of the atmosphere used for assessing the performance of aircraft and aircraft instruments, where pressure and temperature are defined at all heights.

standard deviation (*stats*) The standard deviation of a set of data gives an indication of the amount of dispersion, or scatter, of the members of the set from the measure of central tendency. Its value is the root-mean-square value of the members of the set. The standard deviation is indicated by the Greek letter delta (δ) using the formula

$$\delta = \sqrt{\left\{\frac{\Sigma(x - \bar{x})^2}{n}\right\}}$$

standard form (*maths*) When a number is written in standard form it has one digit to the left of the decimal point and it is multiplied by 10 raised to some power. Thus 6384 written in standard form becomes 6.384×10^3. Similarly, 0.0312 becomes 3.12×10^{-2} in standard form. When a number is written in standard form the first factor is called the *mantissa* and the second factor is called the *exponent*. Thus in the first example 6.384 is the mantissa and 10^3 is the exponent.
standard temperature and pressure (STP) (*sci*) See: *standard atmosphere*.
state (*sci*) The physical form in which matter can exist, being either solid, liquid or gas.
static unbalance (*sci*) A condition in which a rotating mass, when free to rotate, will stop in such a position that the out-of-balance element of the mass is immediately below the axis in the plane of rotation. The condition is corrected by adding to or subtracting from the mass near its rim so that the mass, when free to rotate, can be stopped and remain stationary in any position. See also: *dynamic unbalance*.
star connection (*elec*) Method of connection for three-phase equipment where each phase is connected to a common neutral point, so that the line current is equal to the phase current and the phase voltage is equal to the line voltage multiplied by $\sqrt{3}$. See also: *delta connection*.
statistic (*maths*) The result of the numerical analysis of a set of quantifiable observations.
stator (*elec*) Stationary part of rotating machine, e.g. electric motor, turbine. See: *rotor*.
steady flow (*sci*) Also known as *viscous flow*, it is the flow of fluid particles, that can be considered as groups of molecules, along streamlines so that at any point in the fluid the velocity of the particles is constant or varies with respect to time in a regular manner. Any random motion will only be on a molecular scale.

See: *laminar flow*; *streamline flow*.
steady-flow energy equation (SFEE) (*sci*) Equation applied to *steady-flow processes* based on the *principle of conservation of energy*:

$$\dot{Q} - \dot{W} = \dot{m}\left[(h_2 - h_1) + \frac{1}{2}(c_2^2 - c_1^2) + g(z_2 - z_1)\right]$$

where subscript 2 refers to mass leaving the system, subscript 1 to mass entering the system, h to specific enthalpy, c to the velocity of the fluid, z to the height relative to a datum point and g the acceleration due to gravity.
steady-state error (*sci*) Error between the desired output value and the measured output value after a dynamic system has settled to a constant value.
steam power plant (*sci*) A power plant whose working fluid is steam, comprising a steam boiler where heat is supplied to the system, a steam turbine driving a generator where work is done on the surroundings by the system, a steam condenser where heat is rejected by the system, and a water feed pump where work is done on the system by the surroundings. See also: *Rankine cycle*.
steam tables (*sci*) Thermodynamic data for steam and other fluids set out in tables known as steam tables. They are most useful for working fluids that generally exist as vapours and to which the *gas laws* do not apply. Fluid data is related to pressure and *saturation temperature*, e.g. *specific volume*, specific *internal energy* and *enthalpy*.
steel (*matls*) A family of ferrous materials consisting mainly of iron with small amounts of carbon and manganese. Plain carbon steels have a carbon content ranging between 0.1% and 0.3% for low-carbon steels, 0.4% and 0.8% for medium-carbon steels, 0.9% and 1.4% for high-carbon (tool) steels. In addition, there will be up to 1.5% manganese to combine with any residual sulphur that might be present and generally improve the quality of the steel.

Alloy steels generally have a lower carbon content but contain alloying elements such a nickel, chromium, molybdenum, etc. which impart special properties to the steel either directly or by improving its susceptibility to heat treatment.

stepper motor (*elec*) An electric motor in which the rotor moves through a fixed angle in response to a current pulse from a control source such as a computer. Widely used for the accurate positioning of the elements of small CNC machines and robots.

steradian (*maths*) This is the SI unit of soild angular measurement (symbol: *sr*) subtended at the centre of a sphere by an area on its surface numerically equal to the square of the radius of the sphere. See: *luminous intensity.*

stiction (*sci*) Name sometimes given to *static friction*, the frictional force occurring between two surfaces prior to motion commencing.

stochastic (*elec*) Signals that cannot be described by mathematical equations; also called random signals.

strain (*sci*) Measure of the deformation of a material subjected to an applied load. Assuming Hooke's law applies, linear strain can be expressed as $\varepsilon = x/l$, where l is the original length and x is the change in length. See also: *shear strain.*

strain gauge (*sci*) See: *resistance strain gauge.*

streamline flow (*sci*) See: *laminar flow.*

strength (*sci*) Property of a material that is a measure of its ability to withstand an applied force without breaking or its ability to resist stress. See also: *tensile strength*; *compressive strength*; *shear strength*; *yield strength*. Not to be confused with *toughness* which is the property to resist impact loading.

stress (*sci*) Internal resistance of a material when the shape is changed by the application of an external force. Direct stress, $\sigma = F/A$ where F is the applied force and A is the cross-sectional area of the material perpendicular to the line of action of the force. See also: *shear stress.*

stroke (*mech*) A distance equal to twice the throw of a crank or

eccentric. The distance between the limits of linear movement of the piston of a reciprocating engine, compressor or pump.
strut (*sci*) Bar of framed structure subject to a compressive load. See also: *tie*.
stub arbor (*mech*) A short, single-ended milling-machine arbor: (1) for mounting face mills on the spindle of horizontal milling machines for machining surfaces that are perpendicular to the machine table; (2) for mounting shell end mills on vertical or horizontal milling machines; (3) for mounting cutters normally associated with the horizontal milling machine (e.g. side and face cutters, slitting saws, etc.) on a vertical milling machine.
subroutine (*CAE*) A set of program instructions that can be called up repeatedly and entered into the main body of the program as and when required. It is a useful programming tool, for use when a component design has a number of repeated features such as identical clusters of drilled holes in a particular pattern by a program code. For example (Fanuc controller), N400 M98 P2030 L3 means that at block 400 in the main program the subroutine 2030 is called up and repeated three times. The letter P indicates that the subroutine is required and the letter L indicates that the subroutine is to be repeated.
suction lift (*sci*) **1** Height that a fluid in a reservoir may be raised through a pipe by reducing the pressure in the pipe below that acting on the rest of the surface (usually atmospheric pressure). The pressure acting on the surface then forces the fluid up the pipe. **2** A pump capable of raising fluid up a pipe.
sum (*maths*) The numerical result of adding two or more quantities together.
sump (*Mech*) A reservoir placed below a system into which liquids can flow by gravity for storage. A sump is placed below the crankcase of an internal combustion engine to provide a reservoir for the lubricating oil.
superheat (*sci*) The increase in temperature of a dry *saturated*

vapour at a constant pressure above the *saturation temperature*. The degree of superheat of a vapour is measured as the difference between the temperature of the superheated vapour and the corresponding saturation temperature at the pressure of the vapour. In a steam plant, the steam is superheated to increase its potential energy so that it can be worked expansively and, therefore, more efficiently.
surd (*maths*) Numbers containing a root sign that cannot be expressed exactly as a decimal. For example $\sqrt{2} = 1.4142136\ldots$ and $\sqrt{3} = 1.7320508\ldots$, both continuing indefinitely.
surface hardening (*mech*) See: *case hardening*.
surface texture (*mech*) also referred to as surface finish. When magnified, all surfaces are comprised of 'peaks' and 'troughs'. The smaller the amplitude of these irregularities, the better the surface texture. As shown in Fig. S.13, the surface texture may contain *roughness* and *waviness*. Roughness results from the production process being used and is superimposed upon any waviness. Waviness results from deflection of the machine or workpiece, or from wear in the machine slides. *Lay* is the direction of the predominant surface pattern. Surface texture measurement is usually taken at right angles to the lay. *Sampling length* is the profile length selected for surface texture measurement. *Reference line* is the line chosen by convention to serve for quantitative evaluation of the roughness of the effective profile. The preferred methods of grading surfaces currently employed are the arithmetic mean deviation (R_a), formerly known as the centre-line average (CLA) and the peak-to-valley height index (R_z). The R_a value is used when there is sufficient length of surface to take several sample readings; the R_z value is used when only a short length of surface is available.
surroundings (*sci*) Everything outside a *system* boundary that may be affected by the system.
swages (*mech*) Hand-forging tools used for reducing and finishing round or hexagonal sections, in pairs with half-grooves of dimensions to suit the workpiece. The lower swage is held in the hardie-

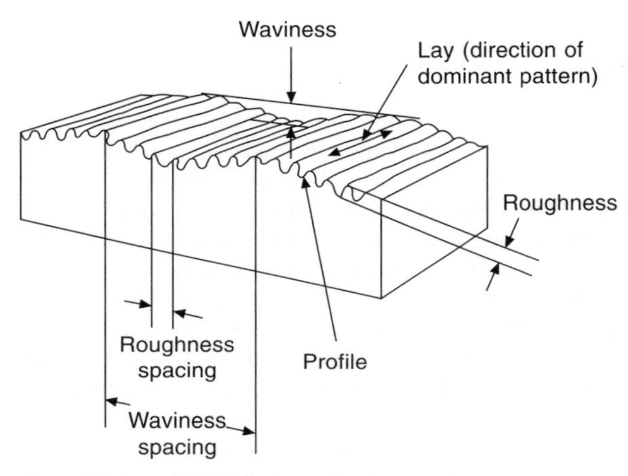

Fig. S.13 Surface Texture Terminology

hole of the anvil and the upper swage is fitted with a handle so that it can be safely held whilst being struck with a hammer.
swept volume (*sci*) The volume through which the piston sweeps in a cylinder of an internal combustion engine. It is calculated as the difference between the maximum cylinder volume when the piston is at the bottom of its stroke and the clearance volume when the piston is at the top of its stroke.
synchronous motor (*elec*) An alternating electric current motor designed to run at a constant speed in synchronization with the supply frquency. Widely used in such devices as electric clocks.
system (*sci*) Collection of connected components or matter under investigation within a boundary defined to simplify problems. Inputs and outputs of the system are shown crossing the boundary.
systematic error (*elec*) Any error of instrumentation or a system that remains constant from one reading to another.

T

tailstock (*mech*) The tailstock of a centre lathe is mounted on the machine bed opposite the headstock. Its position can be adjusted longitudinally on slideways that are separate from but parallel to the carriage slideways so that its accuracy is unaffected by wear. The tailstock carries a barrel with a morse taper bore that is in line (coaxial) with the headstock spindle. The barrel can be wound in and out using a hand wheel and carries the right-hand *centre* when supporting work between centres. *Drills* and *drill-chucks* can also be held in the taper bore. The tailstock can be adjusted to one side for taper turning. See: *centre lathe*.

tally chart (*stats*) A chart used for accurately counting members into their various classes, as shown in Fig. T.1.

Class	Class Mid-point	Tally	Frequency
1 to 5	3	I	1
6 to 10	8	~~IIII~~ I	6
11 to 15	13	~~IIII~~ ~~IIII~~ ~~IIII~~ III	18
16 to 20	18	~~IIII~~ ~~IIII~~ ~~IIII~~ ~~IIII~~ II	22
21 to 25	23	~~IIII~~ ~~IIII~~ ~~IIII~~ IIII	19
26 to 30	28	IIII	4
31 to 35	33	I	4
35 to 40	38	Nil	

Fig. T.1 Tally Chart

tap (*mech*) A tool for cutting an internal thread in a hole, by hand or machine. A hole is drilled with the correct core diameter and then a tap is 'screwed' into the hole by means of a tap wrench. A set of taps comprises a taper tap, a second tap, and then a plug tap. The full set is only needed when tapping to the bottom of a blind hole, otherwise only the first two taps used sequentially are needed.
taper turning attachment (*mech*) Device for taper turning on a centre lathe that causes the tool to move in a path inclined to the common axis of the lathe centres, allowing the centres to remain in alignment. The attachment, which is mounted on the rear of the lathe, carries a slide bar that can be inclined in the horizontal plane so that it is at the required half-angle to the axis of the workpiece. It is connected by a link to the cross-slide. As the carriage traverses along the bed, the cross-slide (and tool) move to or from the work axis following a path parallel to the slide bar of the taper-turning attachment.
Taylor's series (*maths*) Also known as Taylor's theorem. A series expansion of a continuous function, which gives the value of the function for one value of the independent variable in terms of that for another value. This can be expressed as:

$$\mathrm{f}(a + b) = \mathrm{f}(a) + bf'(a) + \frac{b^2}{2!}f''(a) + \ldots$$

Some applications of Taylor's theorem include numerical differentiation, limits and the numerical solution of certain differential equations. See: *Maclaurin's series*.
temperature (*sci*) Property of a system that determines the amount and direction of heat transfer that will take place between the system and its surroundings. See: *Kelvin*; *Celsius*.
temperature coefficient of resistance (*elec*) The resistance of most materials is affected by variations in temperature. Pure metals show a marked increase in resistance with any increase in temperature. Metal alloys show a smaller or even negligible increase in resistance as the temperature increases. Non-metals, including

carbon—the only non-metallic conductor, show a decrease in resistance with any increase in temperature. Therefore the temperature of insulators should not be allowed to rise above the design level. The *temperature coefficient of resistance of a substance is the change in resistance of 1 ohm at 0°C for each degree temperature rise*. Since it is not convenient to work at 0°C, the formula for calculating the temperature effect on resistance is:

$$R_1/R_2 = (1 + \alpha_0\theta_1)/(1 + \alpha_0\theta_2)$$

where R_1 = initial resistance, R_2 = final resistance, α_0 = temperature coefficient of resistance in ohms per ohm per degree celsius, θ_1 = initial temperature, θ_2 = final temperature.

tempering (*matls*) A heat-treatment process following quench hardening for increasing the toughness of steel at the expense of some hardness. The hardened steel is reheated to the appropriate tempering temperature for its intended application and again quenched. For example, knives and edge tools would be tempered at 230°C as hardness rather than toughness is required, whilst an axe head would be tempered at 275°C as a greater degree of toughness is required. Thus the increase in toughness depends upon the tempering temperature reached.

tenon (*mech*) Raised part of a component or a separate tenon block that can be located in a recess or slot (tenon slot) for positioning the component. For example, setting a machine vice parallel to the T-slots of a machine table.

tensile strength (*sci*) The maximum *stress* a material will withstand when loaded in *tension*. Not to be confused with the breaking stress which is lower than the maximum tensile stress (MTS) owing to thinning or 'necking' of the test specimen which reduces its cross-sectional area.

tensile tests (*sci*) Tests carried out on standard samples of a material to investigate the behaviour of the material under different tensile (stretching) loads, from zero to the breaking point. The extension of the sample specimen is measured for each increase in load. The results are then plotted on a graph whose axes will be either load and extension or stress and strain. An example of a stress/strain curve for annealed mild steel is shown in Fig. T.2.

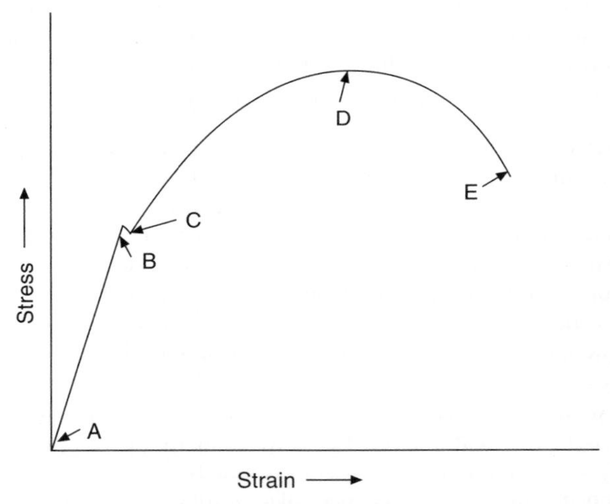

A → B Elastic deformation (Strain is proportional to stress)
B Elastic limit (Limit of proportionality)
C Yield point
C → E Plastic deformation
D Ultimate tensile stress for the material (UTS)
E Breaking stress

Fig. T.2 Typical Tensile Test Curve for Annealed low-carbon (mild) steel

tension (*sci*) When two direct forces acting in opposite directions in the same plane are applied to a material so that they tend to stretch it, the material is said to be *in tension*.

terminal potential difference (*elec*) The measured potential across the terminals of a source of electrical energy when a load is applied to that source. All sources of electrical energy have an *internal resistance* that is in series with the external load resistance. The e.m.f. of the source is measured at the terminals under *no-load* conditions. As soon as a load is connected, the terminal potential falls and the e.m.f. becomes the sum of the potential differences across the internal resistance and the external load. The greater the external load the lower will be the terminal potential difference. This is why the lights of a car dim when the starter is operated.

tesla (*elec*) Unit of magnetic flux density, symbol *T*; $1\ \mathrm{T} = 1\ \mathrm{Wb/m^2}$.

thermal conduction (*sci*) See: *conduction.*

thermal insulation (*sci*) See: *insulation.*

thermistor (*elec*) Semiconductor sensing element used for temperature measurement. Its decrease in resistance is proportional to the rise in temperature.

thermocouple (*sci*) Device comprised of two dissimilar electrical conductors welded together at the ends to form a circuit. It is used as a sensing element to measure temperature. The junction of the two dissimilar metals is called the *hot junction* and the connection to an indicating instrument is called the *cold junction.* If a temperature difference occurs between the two junctions an electric current will flow around the circuit. This is proportional to the temperature difference and can be measured. Usually the cold junction is kept as a reference and the hot junction is applied as the sensing element. See: *pyrometer.*

thermodynamics—first law (*sci*) This states that: 'when a system undergoes a thermodynamic cycle, the net heat supplied from the surroundings to the system is equal to the net work done on the surroundings by the system'. This is a particular form of the *principle of conservation of energy*. For a system that is not a cycle then the *non-flow energy equation* applies as the energy entering a system minus the energy leaving the system is equal to the change of energy in the system.

thermodynamics—second law (*sci*) This states that: 'when a thermodynamic cycle extracts heat from a reservoir, it cannot do an equivalent amount of work on the surroundings', i.e.

$$\Sigma\, \delta W < \Sigma\, \delta Q$$

This has different interpretations resulting from the irreversibility of natural processes and is often written in other forms. For example: 'a thermodynamic cycle cannot transfer heat from a cooler to a hotter body without work being done on the system by the surroundings', or 'work can always be transformed completely and continuously into heat whereas heat can never be transformed completely and continuously into work'.

thermodynamics—third law (*sci*) This states that 'as the temperature of a substance approaches absolute zero the entropy also approaches zero'.

thermodynamic system (*sci*) A system comprised of a region in space containing a quantity of matter under investigation. The boundary of the system is carefully defined and may be fixed or elastic, open or closed. See also: *open system*; *closed system*.

thermodynamics—Zeroth law (*sci*) This states that 'if two bodies are in thermal equilibrium with a third body then all three bodies are in thermal equilibrium'.

Thevenin equivalent circuit (*elec*) Any series–parallel network of resistors, however complicated, can be represented by a voltage source of internal resistance R_o feeding a single equivalent resistor R_{eq}, as shown in Fig. T.3. See: *Norton equivalent circuit*.

threading dial (*mech*) Device fitted to the carriage of a centre lathe comprising a dial connected by a gearwheel to the lead screw for indicating the correct moment in time of engagement of the half-nut when screw cutting. This enables the thread-cutting

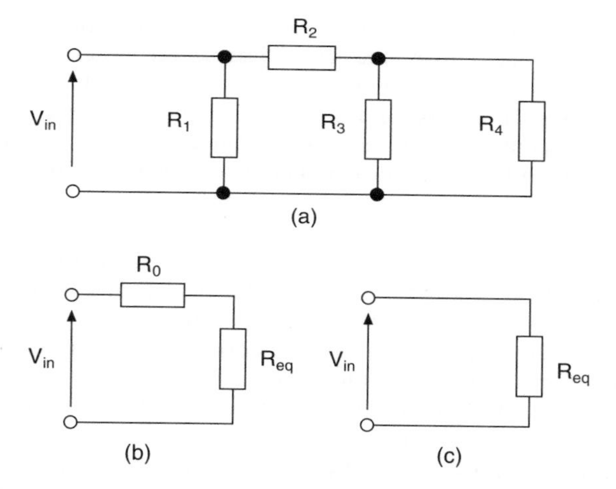

Fig. T.3 (a) Series–parallel network; (b) Thevenin equivalent circuit; (c) Simplified equivalent circuit

tool to follow the helix being cut repeatedly as successively deeper cuts are taken.

thrust bearing (*mech*) A bearing designed to resist an axial load, e.g. the thrust block of a ship transfers the forward thrust from the propeller and shaft to the ship's hull.

tie (*sci*) A member of a framed structure supporting a tensile load.

TIG process (*mech*) (tungsten inert gas) A gas shielded, arc-welding process for producing high-quality welds. The arc is struck between a tungsten electrode and the workpieces being joined. A separate filler rod is used as in oxyacetylene welding. A fusion weld is achieved under a blanket of *argon* or *helium*, or sometimes *carbon dioxide* which is cheaper, but less effective, to prevent atmospheric attack of the hot and molten metal.

time base (*elec*) A line appearing horizontally across the screen of an oscilloscope. It is generated by the *time base generator*, the output of which is applied to the X-plates of the cathode ray tube. When a signal of varying amplitude is applied to the Y-plates, it appears on the screen as a trace of amplitude plotted against time.
time constant (*elec*) For any quantity that varies exponentially with time, the time constant is the time taken for a fractional change of amplitude equal to $100(1 - 1/e) = 63\%$, where e is the exponential constant (the base of natural logarithms). When a capacitor C is charged from a constant voltage source through a resistance R, the time constant is RC. The time constant for the current flowing through an inductance L in series with a resistance R, when fed from a constant voltage source, is L/R.
time-division multiplexing (*elec*) A method of mutiplex transmission that has become available with the advent of digital technology in which the signal is in the form of pulses. The spaces between the pulses are sampled and additional signal pulses are interleaved between them. To extract the required signal at the receiving end a system operating in synchrony with the transmitter is required. See: *multiplexing*.
tin (*matls*) A soft, corrosion-resistant metal (melting point 231.9°C) used to coat mild steel sheet (*tin plate*) and as an alloying element in soft-solders, bronzes and white bearing metals.
tinning (*mech*) The coating of the soldering-iron bit and the joint surfaces of the workpieces with a film of soft-solder prior to making the soldered joint. This ensures a sound joint and uniform penetration of the solder into the joint.
tolerance (*mech*) The *numerical difference* between the upper and lower limits of a dimension. The dimensional range that can be *tolerated* for any feature of a component if it is to function correctly.
tool offset length (*TOL*) (*CAE*) Allows tools of various lengths to be used with a common datum without having to alter the

program. It also allows for the operator to make adjustments for changes in tool length as the result of regrinding when the tools need sharpening. The first tool is 'touched' onto the surface of the work or onto a setting block of known thickness and the *z*-axis read-out is noted. This tool becomes the master tool. The *z*-axis offset is then applied to all the remaining tools to compensate for the difference in length between them and the master tool. The offsets are recorded in the machine's computer under the tool number file. Each time a tool is called up by its T-number the machine automatically selects the correct length offset.

toolmaker's vice (*mech*) A hand-held vice for holding small, delicate work. Also called a pin-vice.

tool-nose radius compensation (*CAE*) The path followed by the program assumes a lathe tool with a perfectly sharp point. In practice, the tool-nose has to have a radius and tool-nose radius compensation automatically allows for this radius. It also allows for changes in radius when tools are changed or reground without having to change the program.

toothed belt (*mech*) A belt drive system in which the teeth moulded into the belt engage with teeth on the pulley wheels thus providing a positive drive that cannot slip like a conventional friction belt drive. Also called a synchronous belt drive. A typical application is the drive between a stepper motor and a lead screw on a small CNC machine. Toothed belts have the advantages of being quieter than chain drives and not requiring lubrication. See: *belt drive*; *Vee belt drive*.

torque (*sci*) Turning moment applied to a shaft calculated as the product of the applied force and the distance from the axis to the force perpendicular to the line of action; unit the newton metre (N m).

torsion equation (*sci*) Equation relating torque and shear stress to material and dimensional quantities of a shaft:

$$\frac{T}{J} = \frac{\tau}{r} = \frac{G\theta}{l}$$

where T is the turning *moment* in N m, J is the *polar second moment of area* in m^4, τ is the maximum *shear stress* at the outer fibres in N/m^2, r is the *radius* to the outer fibres of the *shaft* in m, G is the *Modulus of rigidity* of the shaft material in N/m^2, θ is the angle of twist in *radians*, l is the length of the shaft under torsion in m.

torsional resilience (*sci*) The elastic strain energy stored in a shaft when work is done to twist the shaft, provided the material is not strained beyond the elastic limit, where

$$\text{torsional resilience} = \frac{T\theta}{2} = \frac{\tau^2 V}{4G}$$

(symbols as for the torsion equation).

toughness (*matls*) Ability of a material to withstand impact loads without fracture. Not to be confused with strength. See: *Izod test*; *Charpy test*.

trammel (*mech*) Also called a beam compass. A marking-out tool for striking arcs of large radius and scribing circles of large diameter. The scribing heads are mounted on a wooden or metal beam along which they can be adjusted according to the radius required. One of the scribing heads often has a means of fine adjustment.

transfer function (*maths*) Mathematical expression relating the outputs to the inputs of a system, dynamics generally being represented by *Laplace transforms* rather than differential equations.

transformer (*elec*) *A.C.* machine for altering voltage levels to provide a suitable supply or to minimize losses during transmission, consisting of two windings on a common core, electrically insulated from each other, that rely on mutual inductance for energy transfer. The core is iron for high flux linkage between the coils and is laminated to minimize the eddy currents. The voltage ratio is the same as the *turns ratio* to allow a step up or a step down as required. See: *turns ratio*.

transistor (*elec*) Active semiconductor electronic component in which the current flowing between two of the terminals can be controlled by a smaller current applied to the third terminal. Therefore the device can be used to amplify small signals. The increase in output is achieved by the use of an external power supply. Transistors are the basic component of all electronic circuits from oscillators to computers, and often there are thousands on a single integrated circuit.
transversal (*maths*) Any straight line lying in the same plane and cutting parallel lines.
trapezium (*maths*) Any four-sided plane figure with only two sides that are parallel. Its area = $^1/_2$ × sum of the lengths of the parallel sides × perpendicular distance between them.
trapezoidal rule (*maths*) With reference to Fig. T.4, the

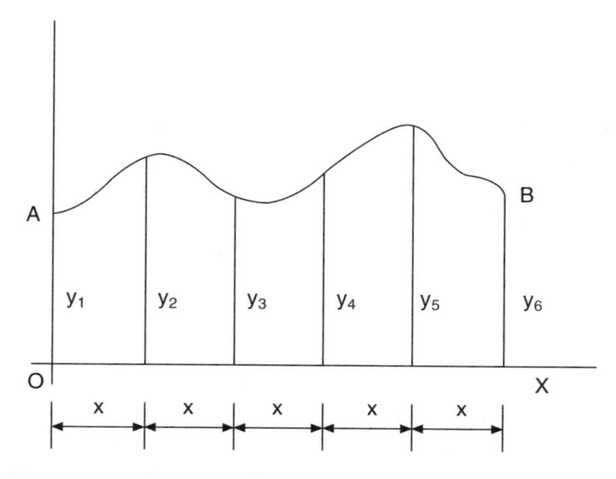

Fig. T.4 Trapezoidal rule

approximate area under the curve AB can be calculated by dividing the base OX into any number of equal intervals (x) (the more intervals, the greater the accuracy). Measure the corresponding ordinates y_1, y_2, y_3, etc., and the area of the figure can be found by applying the formula:

$$\text{area} = x[{}^1/_2(y_1 + y_6) + y_2 + y_3 + y_4 + y_5]$$

Expressed generally, *area* = (*width of interval*) × (*half the sum of the first and last ordinates*) + (*the sum of the remaining ordinates*).

triangle (*maths*) Any figure bounded by three straight lines. The three internal angles always add up to 180°. An *acute-angled triangle* has all its angles less than 90°. A *right-angled triangle* has one angle of 90°, the side opposite the right-angle being called the *hypotenuse*. An *obtuse-angled* triangle has one angle lying between 90° and 180°. An *equilateral triangle* has all three sides of the same length and therefore all its angles are 60°. An *isosceles triangle* has two sides and two corresponding angles equal. A *scalene triangle* is one with unequal angles and, therefore, unequal side lengths. Triangles are said to be *congruent* if they are equal in all respects, i.e. they have the same angles and the same side lengths. Triangles are said to be similar if their angles are the same and if their corresponding side lengths are in proportion.

triangle of forces (*sci*) Vector diagram in the shape of a triangle whereby each side represents one of three coplanar forces. It is useful for representing and solving many engineering problems, e.g. the force acting on the members of a simple frame as shown in Fig. T.5. As the diagram is closed this indicates that the system of forces is in equilibrium and as the vectors act through one point there is no turning effect.

trigonometry (*maths*) Mathematical functions which can be described by right-angled triangles. The side opposite the right angle is called the hypotenuse, the side opposite one other angle,

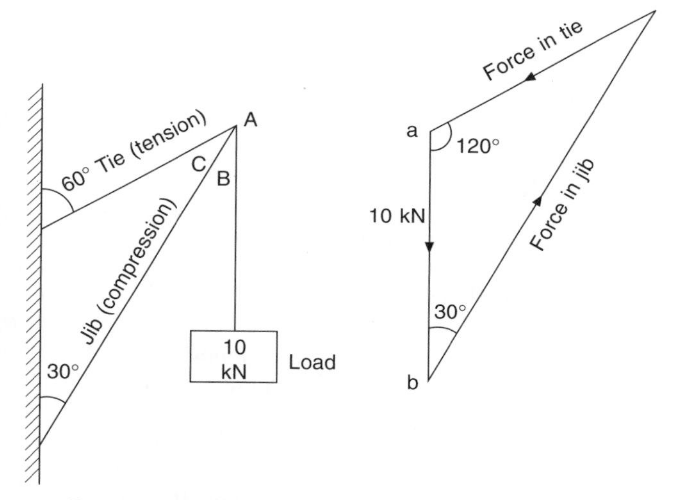

By drawing the triangle of forces to scale, given that ab = 10 kN (load), the forces in the jib and tie can be determined by scaling the sides ac and bc of the force triangle.

Fig. T.5 Triangle of Forces

θ, is called the opposite side and the side between that right angle and the angle θ is called the adjacent side. With reference to Fig. T.6, the relationships are then as follows:

$$\text{tangent of } \theta \text{ or } \tan\theta = \frac{\text{opposite}}{\text{adjacent}}$$

$$\text{cosine of } \theta \text{ or } \cos\theta = \frac{\text{adjacent}}{\text{hypotenuse}}$$

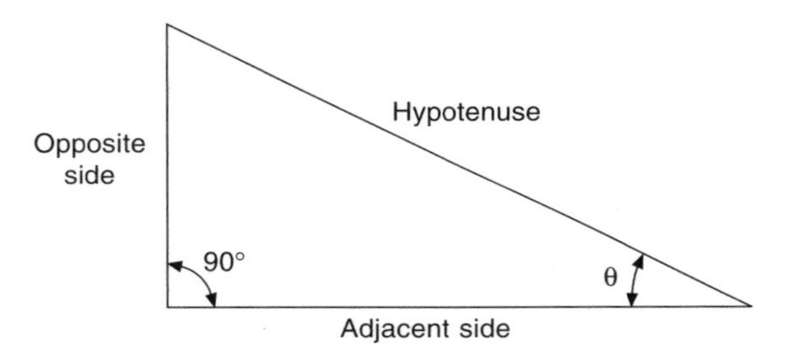

Fig. T.6 Right-angled Triangle (trigonometry)

$$\text{sine of } \theta \text{ or } \sin\theta = \frac{\text{opposite}}{\text{hypotenuse}}$$

The reciprocal of these functions are as follows:

$$\text{cotangent of } \theta \text{ or } \cot\theta = \frac{1}{\tan\theta}$$

$$\text{secant of } \theta \text{ or } \sec\theta = \frac{1}{\cos\theta}$$

$$\text{cosecant of } \theta \text{ or } \operatorname{cosec}\theta = \frac{1}{\sin\theta}$$

tristate logic gate (*elec*) Despite its name, tristate logic gates do NOT have three logic levels. They have the same function as conventional gates except that the output can be switched to a high impedance state using the control input. At control input logic 1 they behave normally, at control input logic 0 the gate output assumes a high impedance state.
true power (*elec*) Power dissipated in an a.c. circuit by resistance only and not by inductance or capacitance. True power = $VI \cos \phi$ where ϕ is the phase angle.
truth table (*elec*) A table for determining the state of the output

of a gate according to the state of its inputs. In drawing up the table, every permutation of input conditions must be shown. Figure T.7 shows a simple OR gate circuit and its corresponding truth table.

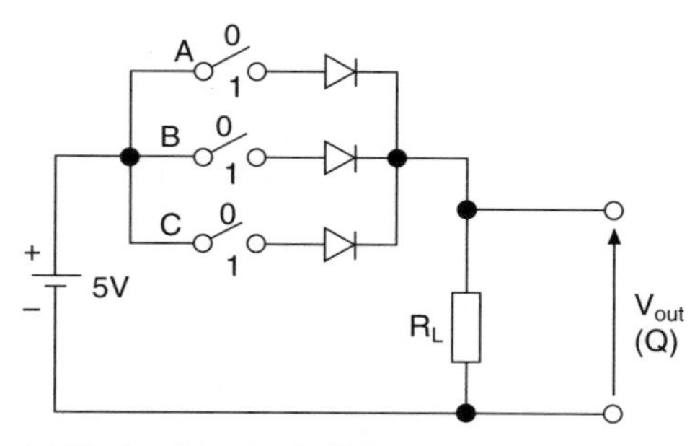

(a) Truth table circuit (OR gate)

A	B	C	Q(V)	Logic value
0	0	0	0	0
0	0	1	4.4	1
0	1	0	4.4	1
0	1	1	4.4	1
1	0	0	4.4	1
1	0	1	4.4	1
1	1	0	4.4	1
1	1	1	4.4	1

(b) Truth table for (a)

Fig. T.7 Truth Table

try-square (*mech*) An instrument for checking the perpendicularity between two surfaces or for marking out lines at right angles to a datum (service) edge. It consists of a heavy stock and a thinner blade. An engineer's square is made of steel and is often used in conjunction with a marking-out table for reference purposes.

TTL (*elec*) Transistor–transistor logic, contemporary integrated circuit technology with typical characteristics of, fan-in = 8, fan-out = 10, propagation delay = 9 ns, noise margin = 0.4 V, power consumption = 40 mW.

turbine (*mech*) **1** *Steam and gas turbines*: machines in which fluids under high temperatures and pressures are made to do mechanical work as they expand. The kinetic energy of the expanding fluid is converted into mechanical work by the rotor of the turbine. Stationary nozzles or vanes direct the expanding, high-velocity working fluid onto blades set in the rotor, causing it to rotate. Turbines are typically used to drive electrical generators. **2** *Water turbine*: the modern development of the water wheel in which the energy of flowing water is made to do mechanical work, usually to drive an alternator in a hydro-electric generating plant. **3** *Wind turbine*: the modern development of the windmill in which the energy of airflow (wind) is made to do mechanical work, usually to drive an electrical generator.

turbulent flow (*sci*) Flow regime of fluid where the motion of the individual fluid particles is chaotic and flow is due to a movement in the general direction of the fluid flow. Turbulent flow occurs at high *Reynold's numbers*. See also: *laminar flow*.

turns ratio (*elec*) Ratio of the turns of one winding of a *transformer* to the other winding. The input winding is called the primary winding and the output winding is called the secondary winding. If the secondary winding has twice the number of turns as the primary winding, then the output voltage will be twice the input voltage and vice versa. In an ideal transformer the output current will be halved in this example. In practice, the output power will be slightly less than the input power because of losses that reduce the efficiency below the ideal 100%.

twist drill (*mech*) A tool that rotates to cut holes. Figure T.8

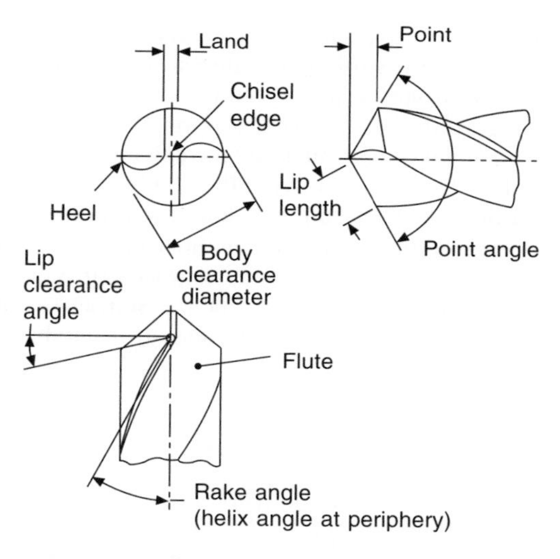

Fig. T.8 Nomenclature of twist drill

shows the nomenclature for the more important features of a twist drill. The lands are narrow strips along the cutting edges of the flutes to provide guidance with minimum friction, and it is the diameter across these that is accurately ground to size. The flutes provide the correct cutting angle and clearance for waste material to flow from the hole. The rake angle (helix angle of the flutes) can vary according to the material to be cut, but the standard angle of a general-purpose 'jobber' drill is 25–30°, the clearance angle is 10–12°, and the point angle is 118°. The diameter of the body tapers slightly (becomes smaller) towards the shank end to provide clearance.

two-stroke cycle (*mech*) An internal combustion engine cycle which requires two strokes to complete and does not have separate induction and exhaust strokes. The piston stroke is longer than an equivalent four-stroke cycle and the piston itself is used to open and close inlet, transfer and exhaust ports. After combustion, the piston moves down the cylinder and the exhaust port is uncovered near the end of the stroke allowing the combustion gases to blow out. Further down the stroke the piston uncovers the transfer port and the air/fuel mixture enters the cylinder and assists in clearing out the remainder of the combustion gases; this is known as *scavenging*. As the piston ascends, the transfer and exhaust ports are covered and the charge is compressed. Near the top of the

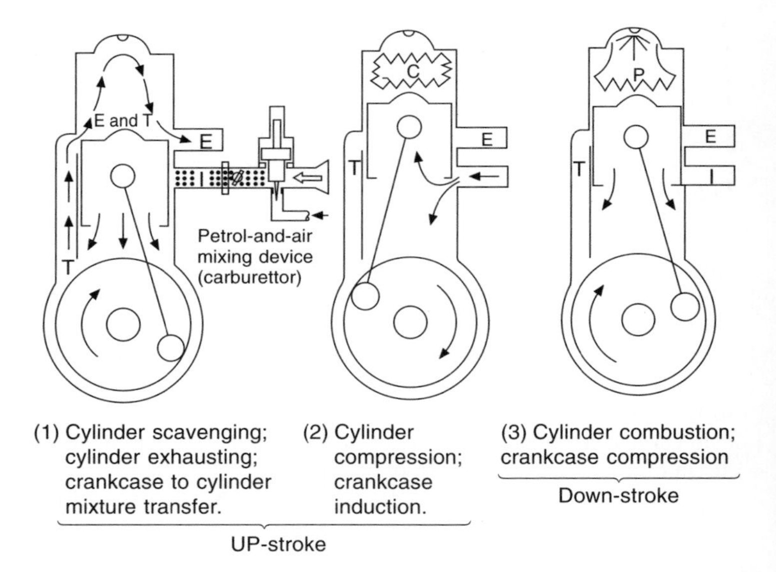

Fig. T.9 Two-stroke Cycle

stroke ignition of the charge takes place and the cycle is repeated. Some engine design arrangements admit the charge initially into the crankcase through a non-return valve or through a port uncovered by the piston skirt as shown in Fig. T.9(2). During the downward motion of the piston the charge is compressed and, as the transfer port is uncovered, the charge passes from the crankcase into the cylinder as shown in Fig. T.9. Two-stroke spark-ignition engines of this type have lubrication oil added to the petrol to allow lubrication through the crankcase. Large marine engines are generally two-stroke compression ignition engines owing to the ease with which the engine may be run directly in reverse without any valves. Some arrangements, however, can be fitted with exhaust valves in the cylinder head and require a cam shift to change moment in the operating cycle when the exhaust valve opens whilst the engine is running in reverse. Two-stroke engines have a higher power-to-weight ratio than an equivalent four-stroke engine.

U

ultimate load (*sci*) Maximum load recorded during a tensile test on a piece of material.
ultimate tensile stress (*sci*) Definition referring to a tensile test on a piece of material, where

$$\text{ultimate tensile stress} = \frac{\text{maximum load}}{\text{original cross} - \text{sectional area}}$$

ultrasonic (*sci*) Pressure waves with a frequency above 20 kHz, which is above the range of normal hearing. Ultrasonic generators convert electrical energy into mechanical energy to generate ultrasonic waves and are used, for example, to test material condition.
unified thread (*mech*) A V-form, screw-thread standard established in 1949 between Britain, Canada and America, with a 60° included angle and radiused roots and crests; the crests of the nut are flat. There are two types: coarse (UNC) and fine (UNF). This thread form, with metric dimensions, is perpetuated in the ISO standard screw thread. See: *ISO screw thread.*
uniformly distributed load (*sci*) Type of load on a beam assumed to be uniformly distributed over a length of the beam rather than acting at a single point. This is usually shown as a bumpy line on space diagrams on top of the beam and is the common way of considering the weight of the beam itself.
unit symbol (*maths*) Symbol of the unit of a scalar quantity which requires a number and a unit symbol, e.g. the unit symbol of a kilogram is ‘kg’. See also the table of SI units in Appendix 4.
unity bracket (*maths*) Mathematical tool for conversion of

units from one form into another. The actual value of a unity bracket is one so anything can be multiplied by a unity bracket as required without affecting the actual value. A unity bracket can be calculated from any relationship. For example 1 inch = 25.4×10^{-3} m, therefore

$$1 = 25.4 \times 10^{-3} \left[\frac{\text{m}}{\text{inch}}\right] = \frac{1}{25.4 \times 10^{-3}} \left[\frac{\text{inch}}{\text{m}}\right]$$

Note that there are two forms of the unity bracket depending upon the conversion required. For example, to convert 15 inches to metres:

$$15 \text{ inches} \times 25.4 \times 10^{-3} \left[\frac{\text{m}}{\text{inch}}\right] = 0.381 \text{ m}$$

The form of the unity bracket to use will have the unit not wanted in the denominator. The use of unity brackets is the best way of ensuring that different units are calculated correctly, particularly in mechanical engineering where units tend to be many and complex.

universal joint (*mech*) coupling to compensate for angular misalignment between rotating shafts in a mechanism.

V

vacuum (*sci*) Space containing no matter but more usually referring to pressures below atmospheric which should more correctly be referred to as partial vacuums.
valves 1 (*elec*) Thermionic devices used for audio-frequency and radio-frequency oscillators and amplifiers. Now largely superseded by solid state devices such as transistors in all but the highest-power applications. **2** (*mech*) Devices for controlling the flow of fluid through an orifice. For example, the opening and closing of a control valve as required, to control the flow of fluid through a pipe.
vane pump (*mech*) A type of pump with a slotted rotor and sliding vanes, which revolves eccentrically in the stator space. The vanes slide in and out creating a variable crescent-shaped volume and providing the pumping action. The higher the number of vanes, the smoother the pumping action.
vapour (*sci*) There is always some vapour above the surface of a liquid as molecules escape from the liquid during the process of evaporation. The hotter the liquid, the greater the agitation of the liquid molecules and the more easily they escape from the surface of the liquid. Therefore the hotter a liquid, the more rapidly it will evaporate. Vapours are not gases and do not obey the gas laws.
vapour pressure (*sci*) The pressure exerted by a vapour, which is the sum of the impact energies of the molecules as they break free from a liquid with increase in temperature and collide with the walls of the vessel in which the vapour is contained.
VAR (*elec*) abbreviation of volt-ampere reactive which is the unit of the reactive component of power. The reactive component of power is the vector difference between the true power and the

apparent power and is equal to VI sinϕ where ϕ is the phase angle.
variance (*stats*) The square of the *standard deviation* of data; a measure of the variability or spread of the data about its mean.
vector quantity (*maths*) Any quantity that has magnitude, sense and direction, e.g. velocity, which is the speed of a body from one point to another in a given direction. See: *scalar quantity*.
vee-belt (*mech*) Drive belt with friction surfaces that have a truncated 'V' section for use with vee-grooved pulley wheels. The standard angle for the 'vee' is 40°. See: *belt drive*; *toothed belt*.
vee-block (*mech*) Devices for supporting shafts and round bars, accurately ground in pairs to ensure that the workpiece is kept parallel with a datum surface whilst marking out or machining. They are rectangular blocks made of *cast iron* or hardened mild steel and having deep V-shaped grooves in two opposing faces.
velocity (*sci*) The rate of change of displacement of a body in a specified direction; velocity is therefore a *vector quantity*, unlike *speed* which is *scalar quantity*. The absolute velocity of a body is its velocity relative to a fixed point on Earth. Relative velocity of a body is its velocity relative to another body, which is the difference of two vector quantities, e.g. the velocity of body B relative to body A is: $v_{BA} = v_B - v_A$.
velocity ratio (*sci*) A commonly used misnomer for the ratio of the distance moved by the effort of a simple machine to the distance moved by the load in the same units. More correctly called the *distance ratio*. See also: *mechanical advantage*.
velocity–time diagram (*sci*) A line graph of *velocity* (y-axis) versus time (x-axis) which is used to solve motion problems and provide a clear picture of events. As the direction of the bodies motion is not usually indicated on the diagrams then, more correctly, they should be called speed–time diagrams. The area under the curve represents the total distance travelled.

vena contracta (*sci*) The point where a jet of fluid has its narrowest cross-sectional area as it flows through an orifice, and where it is less than the cross-sectional area of the orifice itself. See: *coefficient of contraction*.
venturi meter (*sci*) An instrument for measuring volumetric flow rate by measuring the differential pressure across a throat (venturi) in a pipe. The pressure difference is usually measured in terms of the head of fluid in vertical tubes from measuring points in the pipe at both ends of the venturi. The volumetric flow rate is proportional to the loss of pressure head across the venturi.
vernier caliper (*mech*) A measuring instrument for accurately determining the length, thickness, width or inside and outside diameter of a component feature. The instrument consists of a fixed jaw integral with the main beam and a sliding jaw that can be moved along the main beam. The beam carries the fixed scale and the adjustable jaw slide carries the vernier scale, as shown in Fig. V.1(a). The instrument is operated by closing the jaws around the feature to be measured or by opening the jaws when measuring an internal dimension. The distance between the jaws is read by adding the vernier scale reading to the main scale reading, as shown in Fig. V.1(b).
vernier height gauge (*mech*) A precision marking out instrument consisting of a rectangular vertical mast carrying the main scale mounted on a substantial base. The scribing nib and vernier scale are mounted on a head that can slide up and down on the mast and are fitted with a fine adjustment. Measurements are relative to the datum surface upon which the instrument is stood. An example of this instrument is shown in Fig. V.2.
vibrating mass (*sci*) An ideal system to which engineering components can be approximated in many practical, dynamic engineering situations. It consists of a body with a mass of M kg, hanging vertically on a spring which can oscillate up and down, the spring being perfectly elastic, obeying Hooke's law and having

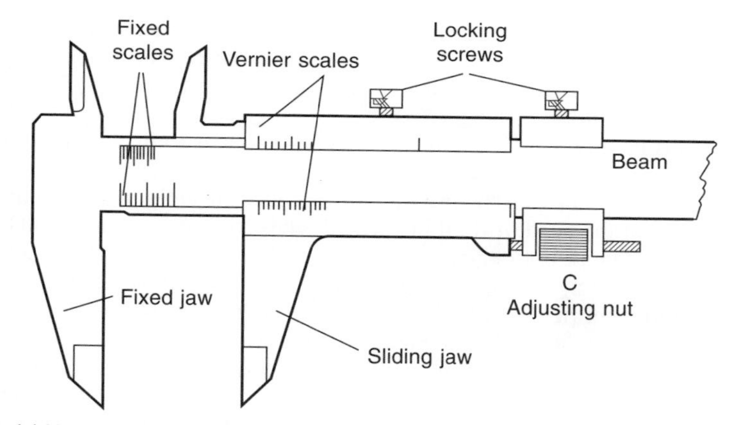

(a) Vernier caliper – general arrangement

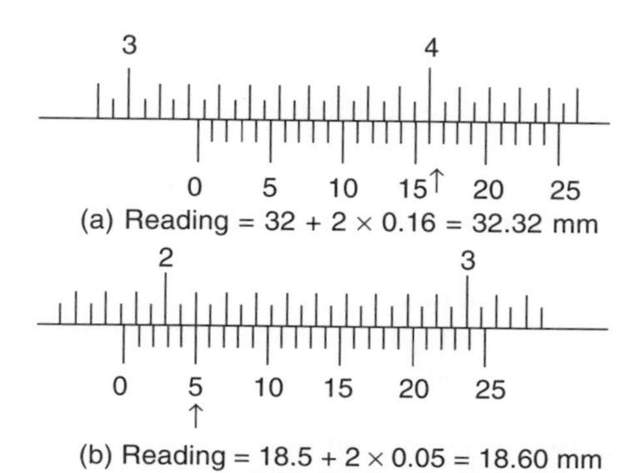

(b) Typical vernier scale readings (Note: not all vernier calipers have the same system. Some have 50 divisions)

Fig. V.1 Vernier Caliper

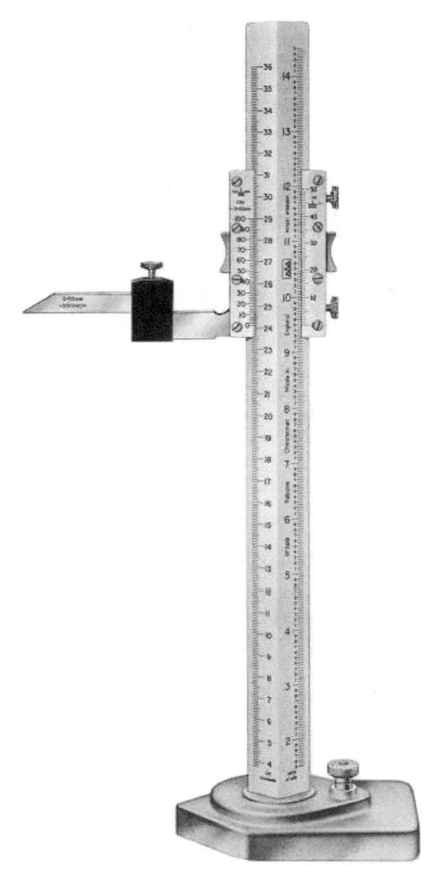

Fig. V.2 Vernier Height Gauge (courtesy of Rabone Chesterman Ltd)

a mass negligible compared with the mass of the body.
vibration (*sci*) A repetitive, periodic change in the displacement of a component, structure or system with respect to some reference point.
vice (*mech*) A device for gripping and holding work; the main type has two parallel jaws, one fixed to the body which is bolted to a bench or machine work table, and the other a sliding jaw adjusted by means of a screw and nut, the screw being rotated by some form of handle. The spindle generally has a *square thread*. The body and slide are *cast iron*, and the replaceable jaws are hardened steel. In a bench vice the jaws are serrated to improve their grip. Soft metal or fibre vice-shoes are available to avoid marking finished surfaces. In a machine vice the jaws are smooth so as not to mark the work. A rapid adjustment lever is sometimes fitted to bench vices for moving the slide quickly in or out for rough positioning.
Vicker's hardness test (*mech*) An indentation test for hard materials in which a square-based diamond pyramid is applied with a standard load to the material under test. The average size of the impression as measured across its diagonals is related to the material hardness by conversion tables. The applied load, which is varied for different materials, has to be stated. This test is considered to be more accurate than the Brinell test for hard materials.
virtual reality (cyberspace) (*comp*) Computer simulation of a real three-dimensional environment that can act on human senses to replicate reality; for example, a headset detects head movement and the view is adjusted automatically, and gloves allow interaction with virtual objects and surfaces. Original systems were developed to train aircraft pilots. Future predictions for virtual reality include sophisticated entertainment systems, training and work assistance.
viscometry (*sci*) The measurement of viscosity by use of a Redwood or similar viscometer. The viscosity is measured in

terms of the number of seconds required for a certain volume of liquid to run out of a cylinder through an orifice under specific conditions.

viscosity (*sci*) The resistance of a fluid to a shearing force and hence to flow and rate of change of shape, measured by the *coefficient of viscosity*, η, with SI unit the pascal-second. *Kinematic viscosity* is sometimes used and this is the viscosity divided by the density of the fluid. Usually, as the temperature of the fluid is increased, the viscocity of the fluid becomes less. Most fluids (Newtonian fluids) can be described by a simple model: two plates a fixed distance apart are filled with the fluid and the upper plate is acted on by a constant force acting parallel to the plates, causing the plates to slide over each other with a constant velocity. The fluid in contact with the fixed plate is static and the fluid in contact with the moving plate is dragged along at the same velocity. Therefore, the velocity of the fluid varies with the distance from the fixed plate, i.e.

$$\eta \propto \frac{F}{\mathrm{d}v/\mathrm{d}y}$$

where F is the applied force, v is the velocity and y is the distance from the plate.

viscous drag (*sci*) An opposing force created by the viscosity of a fluid in a dynamic system. The opposing force resulting from viscous drag is proportional to the velocity of a body moving through the fluid.

voltage (*elec*) Measure of the energy released when a unit of positive charge moves from a higher potential to a lower potential, or the work done in moving charge from a lower potential to a higher potential, quantity symbol V, unit symbol V. It is the voltage between two points of an electric circuit that is the driving force causing the *current* to flow. Also called *electromotive force*. See also: *potential difference*.

voltage multiplier (*elec*) An accurately calibrated resistor connected in series with a moving-coil or moving-iron analogue meter so as to limit the current flow when the meter is used as a voltmeter. In a multi-range meter, various series (multiplier) resistors can be switched into the meter unit so as to give various multiples of the normal full-scale deflection (FSD). See: *moving-coil meter*; *moving-iron meter*.

voltmeter (*elec*) Instrument for measuring the voltage (potential difference) between two points. As voltage is measured between two points then the meter must be connected across (in parallel with) the two points in the circuit and not in line (in series) as with an ammeter. Often a voltmeter is one function of a *multimeter*.

volume (*maths*) Measure of space in three dimensions. It is calculated as the product of three unit lengths, quantity symbol V, SI unit the cubic metre (m^3).

volume of solids (*maths*) The volume of solids can be approximated by a variation of Simpson's rule. In place of the ordinates (y), the cross-sectional area (A) of the solid is used at each interval of length. See: *Simpson's rule*.

volumetric expansion (*sci*) The change in the volume of a mass owing to a change in temperature. This can be predicted by the coefficient of volumetric expansion, γ, units 1/K, where $\delta V = \gamma \times V \times \delta\theta$ where δV is the change in volume and $\delta\theta$ is the change in temperature.

W

Walterizing (*matls*) See: *conversion coatings.*
Washer (*mech*) **1** Annular discs of metal placed between a bolt head and a component or between a nut and a component in order to prevent the surface of the component from being marked as the fastening is tightened up. Such washers may be plain or chamfered. They also provide a smooth and relatively soft surface for the bolt head or nut to bed down onto. **2** Taper washers are used when assembling structural steelwork to prevent the fastening becoming bent as it is tightened onto the tapered surfaces of the beam flanges. **3** Spring washers and serrated washers are placed under a nut to prevent screwed fastenings from working loose.
watt (*elec*) The SI unit of power, symbol W, equivalent to doing work at the rate of one joule per second. In electrical engineering 1 watt = 1 volt × 1 ampere when the load is an ohmic resistor; 1 horse power = 745.7 W.
waveform, average value of, (*maths*) With reference to Fig. W.1 it can be seen that the approximate average or mean value of a waveform (y) can be determined from the expression:

$$y = \text{(area under the curve)/(length of the base OX)}$$

where the area under the curve can be determined by any standard method such as the mid-ordinate rule, Simpson's rule, the trapezium rule, etc. For a sine wave this is zero over a full cycle or 2 × (maximum value)/π for a half-cycle.
wedge angle (*mech*) See: *metal cutting tool angles.*
weld decay (*matls*) A form of pitting corrosion due to the precipitation of carbides which takes place in the heat affected zones adjacent to a weld in nickel chromium alloy steels and,

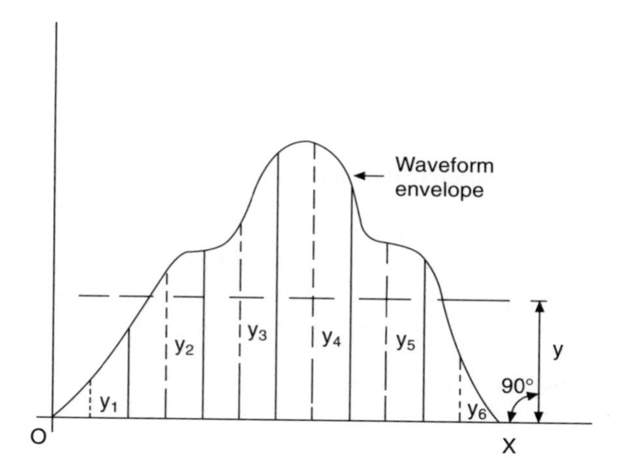

Fig. W.1 Waveform, Average Value of,

particularly, stainless steels. It can be cured by either post-welding heat-treatment of the welded assembly or by using alloys that have been 'proofed' by the addition of miobium or titanium to deter carbide precipitation. Prolonged operation of nickel-chromium alloy and stainless steel components in a temperature range of 650°C to 800°C can have the same effect. Again, the solution is to use proofed alloys.

witness marks (*mech*) When cutting metal to a scribed line it is not possible to see when the line has been 'split'. Therefore, after marking out the workpiece, fine centre punch marks are placed at intervals along the scribed line. When the line is 'split', half the centre punch marks should still be visible as a *witness* that the work has been correctly performed.

word (*CAE*) A 'word' is an associated group of characters that define a complete element of information, e.g. N150. There are two types of word: *dimensional words* and *management words*. See: *dimensional words*; *management words*; *characters*.

word (or letter) address system (*CAE*) In this format system each word commences with a letter character called its *address* and is not dependent on its position in the block (unlike the fixed block format). Therefore instructions that do not have to be changed from a previous block may be omitted from the subsequent blocks until a change becomes necessary. A typical letter address structure (format) taken from a maker's handbook could be:

- Metric: N4 G2 X4,3 Y4,3 Z4,3 I4,3 J4,3 K4,3 F3 S4 T2 M2
- Inch: N4 G2 X3,4 Y3,4 Z3,4 I3,4 J3,4 K3,4 F3 S4 T2 M2

This is interpreted as follows:

- The letter signifies the function of the word.
- The number determines the number of digits that may follow the letter.
- The comma determines the position of the decimal point.

Z

zener diode (*elec*) In an ordinary diode, the avalanche, reverse current effect will destroy the diode (See: *avalanche current*). In a zener diode the depletion layer is deliberately made very thin so that it breaks down at low values of reverse voltage. This limits the reverse, avalanche current to a value that will not destroy the diode. Zener diodes are deliberately operated under reverse bias conditions to provide a known reference voltage for voltage-stabilised power supplies for electronic equipment.

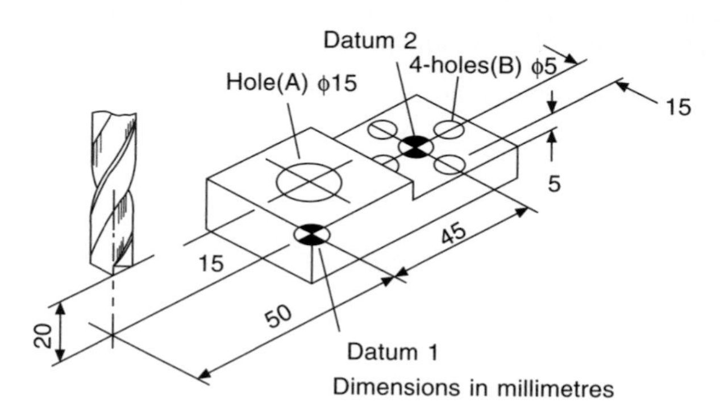

Fig. Z.1 Zero Shift

zero potential (*elec*) A reference potential above or below which other potentials or 'levels' may be related. (1) Theoretical zero potential is said to be at a point in space at an infinite distance from the nearest charged particle (2) Practical zero potential is taken as the potential of the Earth.

zero shift (*CAE*) This is commonly available on most control systems. It allows the program datum to be changed within the program. For the component shown in Fig. Z.1, the program is written to bore hole A using datum 1. Zero shift is then used to move the datum to position 2 as this is more convenient for programming the cluster of holes B about datum 2.

zinc (*matls*) A soft, white metal having low structural strength but good resistance to atmospheric corrosion at ambient temperatures. It has a low melting point of 419.5°C. It is widely used in the construction industry for roofing and flashings, and in engineering and allied industries for anti-corrosion treatments of ferrous metal products. See: *galvanizing* and *sherardizing.* It is also used for sacrificial anodes for the anodic protection of ferrous metal products. See: *Anodic protection.*

Appendix 1

Laplace transforms

The solution of most electrical circuit problems can be reduced ultimately to the solution of different equations. The use of Laplace transforms provides an alternative method for solving linear differential equations.

Definition

The Laplace transform of the function $f(t)$ is defined by the integral

$$\int_0^\infty e^{-st} f(t)\,dt \qquad (1)$$

where s is a parameter assumed to be a real number.

Common notations used for the Laplace transform

There are various commonly used notations for the Laplace transform of $f(t)$ and these include:

(i) $\mathscr{L}\{f(t)\}$ or $\mathscr{L}\{f(t)\}$
(ii) $\mathscr{L}(f)$ or $\mathscr{L}f$
(iii) $\tilde{f}(s)$ or $\tilde{f}(s)$

Also, the letter p is sometimes used instead of s as the parameter. The notation adopted in this book will be $f(t)$ for the original function and $\mathscr{L}\{f(t)\}$ for its Laplace transform. Hence, from equation (1) above,

$$\mathscr{L}\{f(t)\} = \int_0^\infty e^{-st} f(t)\,dt \qquad (2)$$

Table Elementary standard Laplace transforms

Function $f(t)$	Laplace transforms $\{f(t)\} = \int_0^\infty \mathrm{e}^{-st} f(t)\,\mathrm{d}t$
1	$\frac{1}{s}$
k	$\frac{k}{s}$
e^{at}	$\frac{1}{s-a}$
$\sin at$	$\frac{a}{s^2+a^2}$
$\cos at$	$\frac{s}{s^2+a^2}$
t	$\frac{1}{s^2}$
t^2	$\frac{2!}{s^3}$
$t^n (n = 1, 2, 3, \ldots)$	$\frac{n!}{s^{n+1}}$
$\cosh at$	$\frac{s}{s^2-a^2}$
$\sinh at$	$\frac{a}{s^2-a^2}$
$\mathrm{e}^{at} t^n$	$\frac{n!}{(s-a)^{n+1}}$
$\mathrm{e}^{at} \sin \omega t$	$\frac{\omega}{(s-a)^2+\omega^2}$

Function $f(t)$	Laplace transforms $\{f(t)\} = \int_0^\infty e^{-st} f(t)\, dt$
$e^{at} \cos \omega t$	$\dfrac{s-a}{(s-a)^2 + \omega^2}$
$e^{at} \sinh \omega t$	$\dfrac{\omega}{(s-a)^2 - \omega^2}$
$e^{at} \cosh \omega t$	$\dfrac{s-a}{(s-a)^2 - \omega^2}$

Appendix 2

Derivatives

Standard differentials

$y = f(x)$	$\dfrac{dy}{dx}$
ax^n	anx^{n-1}
$\sin ax$	$a \cos ax$
$\cos ax$	$-a \sin ax$
$\tan ax$	$a \sec^2 ax$
e^{kx}	$k\, e^{kx}$
$\ln ax$	$\dfrac{1}{x}$

Differential coefficients of hyperbolic functions

y or $f(x)$	$\dfrac{dy}{dx}$ or $f'(x)$
$\sinh ax$	$a \cosh ax$
$\cosh ax$	$a \sinh ax$
$\tanh ax$	$a\, \text{sech}^2\, ax$
$\text{sech}\, ax$	$-\, a\, \text{sech}\, ax \tanh ax$
$\text{cosech}\, ax$	$-\, a\, \text{cosech}\, ax \coth ax$
$\coth ax$	$-\, a\, \text{cosech}^2\, ax$

Differential coefficients of inverse hyperbolic functions

y or $f(x)$	$\frac{dy}{dx}$ or $f'(x)$
(i) $\operatorname{arsinh} \frac{x}{a}$	$\frac{1}{\sqrt{(x^2 + a^2)}}$
$\operatorname{arsinh} f(x)$	$\frac{f'(x)}{\sqrt{\{[f(x)]^2 + 1\}}}$
(ii) $\operatorname{arcosh} \frac{x}{a}$	$\frac{1}{\sqrt{(x^2 - a^2)}}$
$\operatorname{arcosh} f(x)$	$\frac{f'(x)}{\sqrt{\{[f(x)]^2 - 1\}}}$
(iii) $\operatorname{artanh} \frac{x}{a}$	$\frac{1}{a^2 - x^2}$
$\operatorname{artanh} f(x)$	$\frac{f'(x)}{\{1 - [f(x)]^2\}}$
(iv) $\operatorname{arsech} \frac{x}{a}$	$\frac{-a}{x\sqrt{(a^2 - x^2)}}$
$\operatorname{arsech} f(x)$	$\frac{-f'(x)}{f(x)\sqrt{\{1 - [f(x)]^2\}}}$
(v) $\operatorname{arcosech} \frac{x}{a}$	$\frac{-a}{x\sqrt{(x^2 + a^2)}}$
$\operatorname{arcosech} f(x)$	$\frac{-f'(x)}{f(x)\sqrt{\{[f(x)]^2 + 1\}}}$
(vi) $\operatorname{arcoth} \frac{x}{a}$	$\frac{a}{a^2 - x^2}$
$\operatorname{arcoth} f(x)$	$\frac{f'(x)}{\{1 - [f(x)]^2\}}$

Differential coefficients of inverse trigonometric functions

y or $f(x)$	$\frac{dy}{dx}$ or $f'(x)$
(i) $\arcsin \frac{x}{a}$	$\frac{1}{\sqrt{(a^2 - x^2)}}$
$\arcsin f(x)$	$\frac{f'(x)}{\sqrt{\{1 - [f(x)]^2\}}}$
(ii) $\arccos \frac{x}{a}$	$\frac{-1}{\sqrt{(a^2 - x^2)}}$
$\arccos f(x)$	$\frac{-f'(x)}{\sqrt{\{1 - [f(x)]^2\}}}$
(iii) $\arctan \frac{x}{a}$	$\frac{a}{a^2 + x^2}$
$\arctan f(x)$	$\frac{f'(x)}{\{1 + [f(x)]^2\}}$
(iv) $\text{arcsec} \frac{x}{a}$	$\frac{a}{x\sqrt{(x^2 - a^2)}}$
$\text{arcsec} f(x)$	$\frac{f'(x)}{f(x)\sqrt{\{[f(x)]^2 - 1\}}}$
(v) $\text{arccosec} \frac{x}{a}$	$\frac{-a}{x\sqrt{(x^2 - a^2)}}$
$\text{arccosec} f(x)$	$\frac{-f'(x)}{f(x)\sqrt{\{[f(x)]^2 - 1\}}}$
(vi) $\text{arccot} \frac{x}{a}$	$\frac{-a}{a^2 + x^2}$
$\text{arccot} f(x)$	$\frac{-f'(x)}{\{1 + [f(x)]^2\}}$

Appendix 3

Standard integrals

$y = f(x)$	$\int y \, dx$
ax^n	$\frac{ax^{n+1}}{n+1} + C(n \neq -1)$
$\sin ax$	$-\frac{1}{a} \cos ax + C$
$\cos ax$	$\frac{1}{a} \sin ax + C$
$\sec^2 ax$	$\frac{1}{a} \tan ax + C$
e^{kx}	$\frac{1}{k} e^{kx} + C$
$\frac{1}{x}$	$\ln x + C$

Appendix 4

Système International d'Unités

The International System of Units (abbreviated SI) is commonly called the metric system because it is based upon the metre as the standard unit of length.

The origins of SI go back to Napoleonic France. It is said that Napoleon himself instigated the system, annoyed at being cheated by the merchants who supplied cloth for his army's uniforms. Measurement up to that time was based on 'standards' which were variable to say the least. In England, the 'yard' was the distance between the nose and the end of an outstretched arm, the 'inch' was the width of a person's thumb and a 'foot', just that, the length of a foot. The wealthiest merchants must have all been the smallest people.

The present SI system was internationally adopted in 1960. It is based on seven fundamental units and two supplementary units. All other units are derived from these.

The fundamental units

Unit	Symbol	Dimension	Standard
metre	m	length	1 650 763.73 wavelengths of the orange-red spectral line of krypton-86.
kilogram	kg	mass	Cylinder of platinum–iridium alloy kept by the International Bureau of Weights and Measures in Paris. The only man-made standard remaining.

Unit	Symbol	Dimension	Standard
second	s	time	The time it takes to complete 9 192 631 770 cycles of radiation by a caesium-133 atom.
ampere	A	electric current	The current in each of two parallel conductors that produces an electromagnetic force of 2×10^{-7} N between the conductors when they are 1 m long and spaced 1 m apart.
kelvin	K	temperature	$\frac{1}{273.16}$ of the thermodynamic temperature of the triple point of water. The 'triple point' of a substance is the temperature at which the solid, liquid and vapour phase of the substance are in equilibrium. 0 K is called 'absolute zero'.
mole	mol	amount of substance	The amount of atoms in 0.012 kg of carbon-12.
candela	cd	luminous intensity	The luminous intensity of $\frac{1}{600\,000}$ m^2 of a black body (a perfect radiator) at the temperature of freezing platinum (2045 K).

The supplementary units

Unit	Symbol	Dimension	Standard
radian	rad	plane angle	The angle at the centre of a circle subtended by an arc equal in length to the radius of the circle.
steradian	sr	solid angle	The angle at the centre of a sphere subtended by an area of the surface of the sphere which is equal to the area of a square having sides equal to the radius.

Some of the derived units

Unit	Symbol	Dimension	Comment
square metre	m^2	area	Area enclosed by a square of side 1 m.
cubic metre	m^3	volume	Volume of a cube of side 1 m. The litre is a unit of volume normally used with fluids. It is not an SI unit. 1 m^3 = 1000 l.
metre per second	m/s	velocity	Rate of change of displacement over time.
metre per second squared	m/s^2	acceleration	Rate of change of velocity over time.
newton	N	force	1 N = 1 kg m/s^2. The force that gives a mass of 1 kg an acceleration of 1 ms^{-2}.
pascal	Pa	pressure, stress	1 Pa = 1 N/m^2. The force of 1 N distributed over an area of 1 m^2.

Unit	Symbol	Dimension	Comment
joule	J	work, energy	1 J = 1 Nm = 1 $kg/m^2/s^2$. The amount of energy transferred when the point of application of a force moves a distance of 1 m in the direction of the force.
watt	W	power	1 W = 1 J/s^1. The rate of energy conversion.
hertz	Hz	frequency	1 Hz = 1 cycle/s. The reciprocal of 1 s.
coulomb	C	electric charge	1 C = 1 As. The amount of charge moved by 1 A of current in 1 s.
volt	V	electric potential difference	1 V = 1 J/A s. The potential which converts 1 J of energy in moving 1 C of charge. Potential difference is frequently and incorrectly called 'voltage'.
ohm	Ω	electric resistance	1 Ω = $1V/A^1$. The amount of resistance present when 1 V results in 1 A of current.

There are a number of units called the additional units, which are not part of SI but because they are used so commonly by the scientific and engineering community they cannot be abandoned. They are tolerated within SI, for the present at least.

Appendix 5

Some of the additional units

Unit	Symbol	Dimension	Comment
litre	l	volume (of a fluid)	$1\ \text{l} = 10^{-3}\ \text{m}^3$
gram	g	mass	$1\ \text{g} = 10^{-3}\ \text{kg}$
tonne	t	mass	1 t = 1000 kg
hour	h	time	1 h = 3600 s
minute	min	time	1 min = 60 s
degree celsius	°C	temperature	One degree on the celsius scale is equal to one degree on the kelvin scale.
degree	°	plane angle	$1^\circ = \frac{\pi}{180}$ rad
decibel	dB	power ratio	Obtained by taking 10 $\log_{10}$ of the ratio of two given powers. Used in electronics and audio engineering.

Appendix 6

Some of the prefixes of the preferred standard form

Since SI units form part of a coherent system of units whereby the derived units are all related to the fundamental units, we have to pay the price that some units are inconveniently large and some are inconveniently small. To cope with this, there is a set of prefixes which allows numbers to be expressed in a convenient form. In the preferred standard form the prefixes represent multiples or sub-multiples of 1000 (10^3).

Prefix	Symbol	Exponent of 10	Value
exa	E	10^{18}	1 000 000 000 000 000 000
peta	P	10^{15}	1 000 000 000 000 000
tera	T	10^{12}	1 000 000 000 000
giga	G	10^{9}	1 000 000 000
mega	M	10^{6}	1 000 000
kilo	k	10^{3}	1 000
		10^{0}	1 (no prefix)
milli	m	10^{-3}	0.001
micro	μ	10^{-6}	0.000 001
nano	n	10^{-9}	0.000 000 001
pico	p	10^{-12}	0.000 000 000 001
femto	f	10^{-15}	0.000 000 000 000 001
atto	a	10^{-18}	0.000 000 000 000 000 001

Note the method of separating tens of thousands with a space. Commas should not be used because they can be confused with a decimal point.

It is, in fact, accepted practice in some countries to represent the decimal place with a comma instead of a full stop.

The kilogram is a special case. Because it already contains a prefix and double prefixes are not used, we have to put the prefix in front of gram. For example 1000 kg would not be written as 1 k kg but either as 10^3 kg or 10^6 g.

Confusion sometimes arises with prefixes used with units^2 and units^3. An area expressed as 10 km^2 is an area of:

$$10\ (\text{km})^2 = 10 \times (10^3)^2\ \text{m}^2 = 10 \times 10^6\ \text{m}^2 = 10\ 000\ 000\ \text{m}^2.$$

The use of upper and lower case letters must obey the conventions of SI. Although not preferred, two non-SI prefixes are still used, namely:
centi symbol c 1/100 part, as in centimetre, where 1 cm = 10 mm or 0.01 m deci symbol d 1/10 part, as in decibel, where 1 db = 0.01 bel

Appendix 7

Conversion factors for units

The conversion factors shown below are accurate to five significant figures where FPS is the foot-pound-second system.

FPS to SI units		**SI to FPS units**	
Acceleration			
1 ft/s^2	= 0.304 80 m/s^2	1 m/s^2	= 3.2808 ft/s^2
Angular velocity			
1 rev/min	= 0.104 72 rad/s	1 rad/s	= 9.5493 rev/min
Area			
1 in^2	= 6.4516 cm^2	1 cm^2	= 0.155 00 in^2
1 ft^2	= 0.092 903 m^2	1 m^2	= 10.764 ft^2
1 yd^2	= 0.836 13 m^2	1 m^2	= 1.1960 yd^2
1 acre	= 0.404 69 ha	1 ha	= 2.4711 acre
Density			
1 lb/ft^3	= 16.018 kg/m^3	1 kg/m^3	= 0.062 428 lb/ft^3
Energy			
1 ft pdl	= 0.042 140 J	1 J	= 23.730 ft pdl
1 ft lbf	= 1.355 82 J	1 J	= 0.737 56 ft lbf
1 kW h	= 3.6000 MJ	1 MJ	= 0.277 78 kW h
1 therm	= 0.105 51 GJ	1 GJ	= 9.4781 therm
Force			
1 pdl	= 0.138 26 N	1 N	= 7.2330 pdl
1 lbf	= 4.4482 N	1 N	= 0.224 81 lbf
Length			
1 in	= 2.5400 cm	1 cm	= 0.393 70 in
1 ft	= 0.304 80 m	1 m	= 3.2808 ft

1 yd	= 0.914 40 m	1 m	= 1.0936 yd
1 mi	= 1.6093 km	1 km	= 0.621 37 mi
Mass			
1 oz	= 28.350 g	1 g	= 0.035 274 oz
1 lb	= 0.453 59 kg	1 kg	= 2.2046 lb
1 cwt	= 50.802 kg	1 kg	= 0.019 684 cwt
1 ton	= 1.0161 tonne	1 tonne	= 0.984 21 ton
Moment of force			
1 lbf ft	= 1.3558 N m	1 N m	= 0.737 56 lbf ft

FPS to SI units		**SI to FPS units**	
Plane angle			
1°	= 0.017 45 rad	1 rad	= 57.296
Power			
1 ft lbf/s	= 1.3558 W	1 W	= 0.737 56 ft lbf/s
1 hp	= 0.745 70 kW	1 kW	= 1.3410 hp
Pressure and stress			
1 in Hg	= 33.864 mbar	1 mbar	= 0.029 53 in Hg
1 lbf/in^2	= 6.8948 k Pa	1 kPa	= 0.145 04 lbf/in^2
1 tonf/in^2	= 15.444 N/mm^2	1 N/mm^2	= 0.064 749 tonf/in^2
Specific heat capacity			
1 Btu/(lb°F)	= 4.1868 kJ/(kg°C)	1 kJ/(kg°C)	= 0.238 85 Btu/(lb°F)
Velocity			
1 ft/s	= 0.304 80 m/s	1 m/s	= 3.2808 ft/s
1 mi/h	= 1.6093 km/h	1 km/h	= 0.621 37 mi/h
volume			
1 in^3	= 16.387 cm^3	1 cm^3	= 0.061 024 in^3
1 ft^3	= 0.028 317 m^3	1 m^3	= 35.315 ft^3
1 yd^3	= 0.764 56 m^3	1 m^3	= 1.3080 yd^3
1 pt	= 0.568 26 litre	1 litre	= 1.7598 pt
1 gal	= 4.5461 litre	1 litre	= 0.219 97 gal

Appendix 8

Conversion tables

Fractional sub-divisions of an inch to decimals and to millimetres

in	in	millimetres	in	in	millimetres
$\frac{1}{64}$	0.015625	0.3969	$\frac{39}{64}$	0.609375	15.4781
$\frac{1}{32}$	0.03125	0.7938	$\frac{5}{8}$	0.625	15.875
$\frac{3}{64}$	0.046875	1.1906	$\frac{41}{64}$	0.640625	16.2719
$\frac{1}{16}$	0.0625	1.5875	$\frac{21}{32}$	0.65625	16.6688
$\frac{5}{64}$	0.078125	1.9844	$\frac{43}{64}$	0.671875	17.0656
$\frac{3}{32}$	0.09375	2.3812	$\frac{11}{16}$	0.6875	17.4625
$\frac{7}{64}$	0.109375	2.7781	$\frac{45}{64}$	0.703125	17.8594
$\frac{1}{8}$	0.125	3.175	$\frac{23}{32}$	0.71875	18.2562
$\frac{9}{64}$	0.140625	3.5719	$\frac{47}{64}$	0.734375	18.6531
$\frac{5}{32}$	0.15625	3.9688	$\frac{3}{4}$	0.75	19.05
$\frac{11}{64}$	0.171875	4.3656	$\frac{49}{64}$	0.765625	19.4469
$\frac{3}{16}$	0.1875	4.7625	$\frac{25}{32}$	0.78125	19.8438
$\frac{13}{64}$	0.203125	5.1594	$\frac{51}{64}$	0.796875	20.2406
$\frac{7}{32}$	0.21875	5.5562	$\frac{13}{16}$	0.8125	20.6375
$\frac{15}{64}$	0.234375	5.9531	$\frac{53}{64}$	0.828125	21.0344
$\frac{1}{4}$	0.25	6.35	$\frac{27}{32}$	0.84375	21.4312
$\frac{17}{64}$	0.265625	6.7469	$\frac{55}{64}$	0.859375	21.8281

in	in	millimetres	in	in	millimetres
$\frac{9}{32}$	0.28125	7.1438	$\frac{7}{8}$	0.875	22.225
$\frac{19}{64}$	0.296875	7.5406	$\frac{57}{64}$	0.890625	22.6219
$\frac{5}{16}$	0.3125	7.9375	$\frac{29}{32}$	0.90625	23.0188
$\frac{21}{64}$	0.328125	8.3344	$\frac{59}{64}$	0.921875	23.4156
$\frac{11}{32}$	0.34375	8.7312	$\frac{15}{16}$	0.9375	23.8125
$\frac{23}{64}$	0.359375	9.1281	$\frac{61}{64}$	0.953125	24.2094
$\frac{3}{8}$	0.375	9.525	$\frac{31}{32}$	0.96875	24.6062
$\frac{25}{64}$	0.390625	9.9219	$\frac{63}{64}$	0.984375	25.0031
$\frac{13}{32}$	0.40625	10.3188	1	1	25.4
$\frac{27}{64}$	0.421875	10.7156	2	2	50.800
$\frac{7}{16}$	0.4375	11.1125	3	3	76.200
$\frac{29}{64}$	0.453125	11.5094	4	4	101.600
$\frac{15}{32}$	0.46875	11.9062	5	5	127.000
$\frac{31}{64}$	0.484375	12.3031	6	6	152.400
$\frac{1}{2}$	0.5	12.7	7	7	177.800
$\frac{33}{64}$	0.515625	13.0969	8	8	203.200
$\frac{17}{32}$	0.53125	13.4938			
$\frac{35}{64}$	0.546875	13.8906	9	9	228.600
$\frac{9}{16}$	0.5625	14.2875	10	10	254.000
$\frac{37}{64}$	0.578125	14.6844	11	11	279.400
$\frac{19}{32}$	0.59375	15.0812	12	12	304.800

Millimetres to inches based on 1 inch = 25.4 millimetres

mm	0	1	2	3	4	5	6	7	8	9
	in	in	in	in	in	in	in	in	in	in
–	–	0.03937	0.07874	0.11811	0.15748	0.19685	0.23622	0.27559	0.31496	0.35433
10	0.39370	0.43307	0.47244	0.51181	0.55118	0.59055	0.62992	0.66929	0.70866	0.74803
20	0.78740	0.82677	0.86614	0.90551	0.94488	0.98425	1.02362	1.06299	1.10236	1.14173
30	1.18110	1.22047	1.25984	1.29921	1.33858	1.37795	1.41732	1.45669	1.49606	1.53543
40	1.57480	1.61417	1.65354	1.69291	1.73228	1.77165	1.81102	1.85039	1.88976	1.92913
50	1.96850	2.00787	2.04724	2.08661	2.12598	2.16535	2.20472	2.24409	2.28346	2.32283
60	2.36220	2.40157	2.44094	2.48031	2.51969	2.55906	2.59843	2.63780	2.67717	2.71654
70	2.75591	2.79528	2.83465	2.87402	2.91339	2.95276	2.99213	3.03150	3.07087	3.11024
80	3.14961	3.18898	3.22835	3.26772	3.30709	3.34646	3.38583	3.42520	3.46457	3.50394
90	3.54331	3.58268	3.62205	3.66142	3.70079	3.74016	3.77953	3.81890	3.85827	3.89764
100	3.93701	3.97638	4.01575	4.05512	4.09449	4.13386	4.17323	4.21260	4.25197	4.29134
10	4.33071	4.37008	4.40945	4.44882	4.48819	4.52756	4.56693	4.60630	4.64567	4.68504
20	4.72441	4.76378	4.80315	4.84252	4.88189	4.92126	4.96063	5.0000	5.0394	5.0787
30	5.1181	5.1575	5.1969	5.2362	5.2756	5.3150	5.3543	5.3937	5.4331	5.4724
40	5.5118	5.5512	5.5906	5.6299	5.6693	5.7087	5.7480	5.7874	5.8268	5.8661
50	5.9055	5.9449	5.9843	6.0236	6.0630	6.1024	6.1417	6.1811	6.2205	6.2598
60	6.2992	6.3386	6.3780	6.4173	4.4567	6.4961	6.5354	6.5748	6.6142	6.6535
70	6.6929	6.7323	6.7717	6.8110	6.8504	6.8898	6.9291	6.9685	7.0079	7.0472
80	7.0866	7.1260	7.1654	7.2047	7.2441	7.2835	7.3228	7.3622	7.4016	7.4409
90	7.4803	7.5197	7.5591	7.5984	7.6378	7.6772	7.7165	7.7559	7.7953	7.8346
200	7.8740	7.9134	7.9528	7.9921	8.0315	8.0709	8.1102	8.1496	8.1890	8.2283
10	8.2677	8.3071	8.3465	8.3858	8.4252	8.4646	8.5039	8.5433	8.5827	8.6220
20	8.6614	8.7008	8.7402	8.7795	8.8189	8.8583	8.8976	8.9370	8.9764	9.0157
30	9.0551	9.0945	9.1339	9.1732	9.2126	9.2520	9.2913	9.3307	9.3701	9.4094
40	9.4488	9.4882	9.5276	9.5669	9.6063	9.6457	9.6850	9.7244	9.7638	9.8031
50	9.8425	9.8819	9.9213	9.9606	10.0000	10.0394	10.0787	10.1181	10.1575	10.1969
60	10.2362	10.2756	10.3150	10.3543	10.3937	10.4331	10.4724	10.5118	10.5512	10.5906
70	10.6299	10.6693	10.7087	10.7480	10.7874	10.8268	10.8661	10.9055	10.9449	10.9843
80	11.0236	11.0630	11.1024	11.1417	11.1811	11.2205	11.2598	11.2992	11.3386	11.3780
90	11.4173	11.4567	11.4961	11.5354	11.5748	11.6142	11.6535	11.6929	11.7323	11.7717
300	11.8110	11.8504	11.8898	11.9291	11.9685	12.0079	12.0472	12.0866	12.1260	12.1654
10	12.2047	12.2441	12.2835	12.3228	12.3622	12.4016	12.4409	12.4803	12.5197	12.5591
20	12.5984	12.6378	12.6772	12.7165	12.7559	12.7953	12.8346	12.8740	12.9134	12.9528
30	12.9921	13.0315	13.0709	13.1102	13.1496	13.1890	13.2283	13.2677	13.3071	13.3465
40	13.3858	13.4252	13.4646	13.5039	13.5433	13.5827	13.6220	13.6614	13.7008	13.7402
50	13.7795	13.8189	13.8583	13.8976	13.9370	13.9764	14.0157	14.0551	14.0945	14.1339
60	14.1732	14.2126	14.2520	14.2913	14.3307	14.3701	14.4094	14.4488	14.4882	14.5276
70	14.5669	14.6063	14.6457	14.6850	14.7244	14.7638	14.8031	14.8425	14.8819	14.9213

mm	0	1	2	3	4	5	6	7	8	9
80	14.9606	15.0000	15.0394	15.0787	15.1181	15.1575	15.1969	15.2362	15.2756	15.3150
90	15.3543	15.3937	15.4331	15.4724	15.5118	15.5512	15.5906	15.6299	15.6693	15.7087
400	15.7480	15.7874	15.8268	15.8661	15.9055	15.9449	15.9843	16.0236	16.0630	16.1024
10	16.1417	16.1811	16.2205	16.2598	16.2992	163386	16.3780	16.4173	16.4567	16.4961
20	16.5354	16.5748	16.6142	16.6535	16.6929	16.7323	16.7717	16.8110	16.8504	16.8898
30	16.9291	16.9685	17.0079	17.0472	17.0866	17.1260	17.1654	17.2047	17.2441	17.2835
40	17.3228	17.3622	17.4016	17.4409	17.4803	17.5197	17.5591	17.5984	17.6378	17.6772
50	17.7165	17.7559	17.7953	17.8346	17.8740	17.9134	17.9528	17.9921	18.0315	18.0709
60	18.1102	18.1496	18.1890	18.2283	18.2677	18.3071	18.3465	18.3858	18.4252	18.4646
70	18.5039	18.5433	18.5827	18.6220	18.6614	18.7008	18.7402	18.7795	18.8189	18.8583
80	18.8976	18.9370	18.9764	19.0157	19.0551	19.0945	19.1339	19.1732	19.2126	19.2520
90	19.2913	19.3307	19.3701	19.4094	19.4488	19.4882	19.5276	19.5669	19.6063	19.6457
500	19.6850	19.7244	19.7638	19.8031	19.8425	19.8819	19.9213	19.9606	20.0000	20.0394

For a full list of conversions refer to BS 350, Parts 1 and 2.